101 KEY IDEAS

PHYSICS

Jim Breithaupt

TEACH YOURSELF BOOKS

For UK order queries: please contact Bookpoint Ltd, 78 Milton Park, Abingdon, Oxon OX14 4TD. Telephone: (44) 01235 400414, Fax: (44) 01235 400454. Lines are open from 9.00–6.00, Monday to Saturday, with a 24-hour message answering service. Email address: orders@bookpoint.co.uk

For U.S.A. & Canada order queries: please contact NTC/Contemporary Publishing, 4255 West Touhy Avenue, Lincolnwood, Illinois 60646–1975, U.S.A. Telephone: (847) 679 5500, Fax: (847) 679 2494.

Long renowned as the authoritative source for self-guided learning – with more than 30 million copies sold worldwide – the *Teach Yourself* series includes over 200 titles in the fields of languages, crafts, hobbies, business and education.

British Library Cataloguing in Publication Data
A catalogue record for this title is available from The British Library.

Library of Congress Catalog Card Number: On file

First published in U.S.A. 2000 by Hodder Headline Plc, 338 Euston Road, London, NW1 3BH.

First published in US 2000 by NTC/Contemporary Publishing, 4255 West Touhy Avenue, Lincolnwood (Chicago), Illinois 60646–1975 U.S.A.

The 'Teach Yourself' name and logo are registered trade marks of Hodder & Stoughton Ltd.

Typeset by Transet Limited, Coventry, England.
Printed in Great Britain for Hodder & Stoughton Educational, a division of Hodder Headline Plc, 338 Euston Road, London NW1 3BH by Cox & Wyman Ltd, Reading, Berkshire.

Impression number	10 9 8 7 6 5 4 3 2 1
Year	2005 2004 2003 2002 2001

Contents

Introduction

Welcome to the **Teach Yourself 101 Key Ideas** series. We hope that you will find both this book and others in the series to be useful, interesting and informative. The purpose of the series is to provide an introduction to a wide range of subjects, in a way that is entertaining and easy to absorb.

Each book contains 101 short accounts of key ideas or terms which are regarded as central to that subject. The accounts are presented in alphabetical order for ease of reference. All of the books in the series are written in order to be meaningful whether or not you have previous knowledge of the subject. They will be useful to you whether you are a general reader, are on a pre-university course, or have just started at university.

We have designed the series to be a combination of a text book and a dictionary. We felt that many text books are too long for easy reference, while the entries in dictionaries are often too short to provide sufficient detail. The **Teach Yourself 101 Key Ideas** series gives the best of both worlds! Here are books that you do not have to read cover to cover, or in any set order. Dip into them when you need to know the meaning of a term, and you will find a short, but comprehensive account which will be of real help with those essays and assignments. The terms are described in a straightforward way with a careful selection of academic words thrown in for good measure!

So if you need a quick and inexpensive introduction to a subject, **Teach Yourself 101 Key Ideas** is for you. And incidentally, if you have any suggestions about this book or the series, do let us know. It would be great to hear from you.

Best wishes with your studies!

Paul Oliver
Series Editor

Preface

This book is intended for those without a background in physics to read and learn about key ideas in physics. The latest ideas of physics such as antimatter, quarks, and superconductivity will undoubtedly form the foundations of engineering and technology in the future. Well-established ideas in physics developed centuries ago continue to provide the principles that underpin other branches of science and form the foundations of communications, computing, engineering and technology. Many of the challenges of the new century are likely to be solved by applying our present knowledge and ideas of physics to new discoveries in materials science. For example, cheaper and more efficient solar cells to generate low-cost electricity would raise living standards in under-developed countries enormously. Physics has always been at the frontiers of science and key discoveries in physics such as X-rays and transistors have provided benefits for everyone.

Knowledge and understanding of physics allows non-specialists to evaluate the impact of new technologies and to fit into place important new discoveries in science. This book provides a concise account of the key ideas of physics in accessible and readable format. The 101 key ideas in this book are presented alphabetically, with diagrams where appropriate and cross-references where relevant. The key ideas cover the amazing new discoveries at the frontiers of the subject such as antimatter and quarks and also the essential ideas and facts for beginners who are seeking an introduction to the subject which is essentially descriptive and non-mathematical.

Activation Processes

An activation process is any process where a particle needs to use a certain amount of energy to go through the process. For example, to escape from an uncharged metal surface, a conduction electron must do a minimum amount of work which is referred to as the work function of the surface. Another example is a chemical reaction where two particles need a certain amount of energy to react. Where a large number of particles are present, there is a continuous spread of kinetic energy among the individual particles, ranging from no particles at zero energy to a peak number of particles at a particular energy to a negligible number at very high energies. In an activation process, only a certain fraction of the total number of particles have enough kinetic energy to undergo the process. The distribution of kinetic energy depends on temperature; the higher the temperature, the greater the mean kinetic energy of the particles. Thus the fraction of particles with sufficient kinetic energy to undergo the process increases as the temperature increases. This is why raising the temperature of the reagents in a chemical reaction speeds the reaction up.

The energy which a particle needs in an activation process is like an energy barrier that prevents particles with insufficient energy going through the process. For a large number of particles at temperature T, the mean kinetic energy of the particles is approximately equal to kT, where k is a constant known as the Boltzmann constant. The number of particles with sufficient kinetic energy able to move across an energy barrier E depends on the ratio E / kT since this ratio is a measure of the energy barrier in relation to the mean kinetic energy of the particles. If the temperature is increased, the energy barrier in relation to the mean kinetic energy of the particles is reduced and therefore more particles can 'overcome' the barrier. Evaporation is another example of an activation process. To escape from a solid or a liquid, a molecule must use a certain amount of energy to overcome the attraction of the other molecules in the solid or liquid. Hence the rate of evaporation is increased by raising the temperature of the substance.

see also...

Ideal Gases; Temperature

Alternating Current

An alternating current is an electric current that repeatedly reverses its direction, usually at a constant frequency. An alternating potential difference (p.d.) or voltage between two points in a circuit causes charge carriers between the two points to repeatedly reverse direction.

★ The waveform of an alternating current or potential difference is a graph of the current or p.d. on the y-axis against time on the x-axis. The waveform for alternating current from the mains is shown below. Currents and p.d.s in circuits connected to the mains, either directly or via transformers, are always sinusoidal.

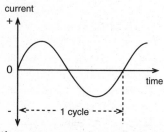

Alternating current

★ The peak value of an alternating current or potential difference is the maximum value of the current or p.d. in either direction. In one complete cycle, the current reverses from its peak value in one direction to its peak value in the opposite direction and back again.

★ The frequency of an alternating current is the number of complete cycles per second. The unit of frequency is the hertz (Hz), equal to 1 complete cycle per second.

★ The root mean square (r.m.s) value of an alternating current or p.d. is the value of the direct current that delivers the same power as the alternating current (or p.d.) to a given resistor.

For a sinusoidal current or p.d., its r.m.s value =

$$\frac{\text{its peak value}}{\sqrt{2}}$$

For example, the r.m.s. value of the mains is 230 V, which means that a mains heater connected to a 230 V d.c. supply would dissipate the same average power as if it was connected to the mains.

> ### see also...
> *Electromagnetic Induction;*
> *Potential Difference and Power*

Antimatter

Matter consists of particles and antimatter consists of antiparticles. For every known type of particle, there is a corresponding type of antiparticle. An antiparticle has a rest mass equal and opposite to the rest mass of its particle counterpart. It has an equal and opposite charge to its particle counterpart if its particle counterpart is charged. The first antiparticle to be discovered was the positron, which is the antiparticle of the electron.

An antiparticle and a corresponding particle can be produced by two methods:

1 A photon of high energy radiation creates a particle and a corresponding antiparticle. The photon ceases to exist as a result. Its energy is converted into matter. e.g. high energy photon → electron + positron.

2 Two particles collide with each other at a speed approaching the speed of light and create another particle and its corresponding antiparticle. Some of the collision energy is converted into matter. e.g. proton + proton → proton + proton + proton + antiproton.

For a high energy photon to produce a particle and its antiparticle, the photon energy (hf) must be greater than or equal to the total **rest energy** of the particle and antiparticle (which is equal to $2m_0c^2$, where m_0 is the rest mass of the particle). A particle and its corresponding antiparticle annihilate each other when they collide, creating two photons in the process.

Galaxies consist of matter not antimatter. Astronomers believe that the Universe was created in a massive explosion about 12 billion years ago. The energy from the Big Bang is thought to have created particles and antiparticles. Many more particles than antiparticles must have been left as the Universe cooled after the Big Bang. The presence of more particles than antiparticles soon after the Big Bang ensured that all the antiparticles were annihilated by particles to form photons. The cause of the asymmetry is thought to be due to the decay of a certain type of high energy antiparticle being slightly more likely than the decay of the corresponding particle.

see also...

Big Bang; Particle Interactions; Photon

Atoms and Molecules

An atom is the smallest particle of an element that is characteristic of the element. An element is a substance that cannot be broken down into other elements. An atom consists of a positively charged nucleus surrounded by negatively charged electrons. The nucleus is composed of protons, which each carry an equal and opposite charge to the electron, and neutrons, which are uncharged. The mass of a proton is almost of the same as a mass of a neutron. The electron's mass is much less than the mass of a proton or neutron.

The electrons of an atom occupy 'shells' that surround the nucleus. The energy of an electron in a shell is constant. Zero energy for an electron in an atom corresponds to the electron outside the atom. Thus the energy of an electron in an atom is negative. The further a shell is from the nucleus, the higher the energy of the electrons in a shell. Each shell can hold a certain maximum number of electrons. The electrons in an atom are normally in the lowest possible energy state, corresponding to the shells occupied from the innermost shell outwards. Atoms join together to form molecules as a result of bonds formed between atoms due to the interaction of their outermost electrons. Each type of atom is represented by ^A_ZX where X is the chemical symbol of the element, Z is its proton number which is referred to as its atomic number, and A is its mass number which is the number of protons and neutrons in the nucleus.

Isotopes are atoms of an element that have different numbers of neutrons in the nucleus. For example, the three isotopes of hydrogen are ^1_1H which consists of 1 proton and 1 electron, ^2_1H which consists of 1 proton and 1 neutron as the nucleus and 1 electron, and ^3_1H which has one more neutron than ^2_1H in its nucleus. The isotopes of an element have different physical properties because each type of atom has a different mass due to the differing number of neutrons. However, because all the atoms of an element have the same number of protons, they all have the same number of electrons and hence the same chemical properties.

see also...

Energy Levels in Atoms; Types of Bonds

Big Bang

The Big Bang theory of the Universe is that the Universe was created in a massive explosion from a point when space, time and matter was created. This event is thought to have occurred about 12 billion years ago. Galaxies formed and moved away from each other as the Universe continued to expand and continues to expand. The distant galaxies are known to be rushing away from each other at speeds approaching the speed of light.

The Big Bang theory originated from the discovery in 1929 by American astronomer Edwin Hubble that distant galaxies are receding at speeds in proportion to their distances. This relationship is known as Hubble's Law and is usually written as:

speed of recession $v = Hd$ for a receding galaxy at distance d, where H is a constant of proportionality known as the Hubble constant.

Although Hubble's Law is explained by the idea that the Universe is expanding, the Big Bang theory was not accepted until the discovery of **cosmic background microwave radiation** in 1965. This radiation was found by Arno Penzias and Robert Wilson who were testing an aerial system to detect radio signals from satellites. They discovered background radiation in the microwave region of the electromagnetic spectrum coming from all directions in space. They concluded this radiation was emitted by matter when the Universe became transparent, shortly after the Big Bang.

Before this discovery, many astronomers favoured the Steady State theory of the Universe which supposed that the expansion of the Universe is due to matter created in space between the galaxies which move away from each other as a result. The Steady State theory was discarded as it cannot explain the presence of cosmic microwave radiation from all directions in space. The Big Bang theory also provides an explanation of why there is three times as much hydrogen as helium in the Universe.

see also...

*Electromagnetic Waves;
Hubble's Law*

Black Hole

Nothing can escape from a black hole, not even light. A black hole is a perfect absorber of all types of electro-magnetic radiation (or any other form of radiation), just as a black surface is a perfect absorber of visible light. The idea was first thought up by John Michell in 1783 although the term 'black hole' is of much later origin as it was coined by the American physicist, John Wheeler. In 1916 Albert Einstein predicted in his General Theory of Relativity that a strong gravitational field distorts space and time and bends light. Einstein calculated that light grazing the Sun from a star was deflected by a thousandth of half a degree due to the Sun's gravity. This prediction was confirmed by Sir Arthur Eddington in 1919 who led an expedition to South America to photograph stars that appeared close to the Sun during a total solar eclipse. Eddington found that the images of these stars on the eclipse photographs were displaced from their normal positions by exactly the amount predicted by Einstein. The modern theory of black holes was started by Karl Schwarzschild who used Einstein's theory to prove that an object with a sufficiently strong gravitational field would prevent light from escaping. Schwarzschild showed that such an object is surrounded by an **event horizon**, a spherical envelope surrounding the object from which nothing inside can escape. Any object falling through it would disappear forever. Its radius is known as the Schwarzschild radius. For a black hole of mass M, its Schwarzschild radius = $2\ GM/c^2$, where G is the constant of gravitation from Newton's theory of gravity and c is the speed of light. The Earth would need to be compressed to less than 18 mm in diameter to become a black hole. Evidence for black holes has been obtained by astronomers. The galaxy M87 is rotating so fast that there is thought to be a massive black hole at its centre. The X-ray source, Cygnus X1, is a binary system consisting of a supergiant star accompanied by a very dense invisible star which may be a black hole attracting matter from its companion.

see also...

General Relativity; Gravitational Fields 1 and 2

Capacitance

A capacitor is any device that can be used to store charge. The capacitance of a capacitor is defined as the charge stored by the device per unit potential difference applied to the device. For a capacitor of capacitance C charged to a potential difference V, the charge stored $Q = CV$. The unit of capacitance is the farad (symbol F) which is equal to 1 coulomb per volt. Typical capacitors used in circuits range from at most 0.001 F down to a fraction of a millionth of a farad. Capacitance values are often expressed in microfarads (μF) where $1\ \mu F = 10^{-6}$ F.

The simplest form of a capacitor consists of two insulated conducting plates facing each other. When a battery is connected to the plates, electrons flow off one plate onto the other. One plate becomes negatively charged because it gains electrons while the other becomes positively charged because it loses electrons. The two plates gain equal and opposite amounts of charge. The charge stored is defined as the amount of charge on either plate.

Energy is stored in a capacitor when it is charged. This energy is released when the capacitor is discharged. For example, if a charged capacitor is discharged through a torch bulb, the electrons from the negative plate of the capacitor flow through the torch bulb onto the positive plate. This discharge current may be large enough to light the torch bulb briefly. For a capacitor of capacitance C charged to a potential difference V, the energy stored $= \frac{1}{2}CV^2$.

Capacitors are used in time-delay circuits, in tuning circuits, in filter circuits and in power supply circuits. The growth and decay of currents, charge and p.d. in a d.c. circuit such as a time-delay circuit is controlled by a capacitor in series with a resistor and a switch. The capacitor discharges at a rate that depends on its capacitance C and on the resistance R of the resistor. The time constant RC is the time taken by the current to fall to 37% of its initial value when the capacitor is charging or discharging in a d.c. circuit.

see also...

Potential Difference and Power

Charge and Current

★ An electric current is a flow of charge. In a metal, the charge is carried by electrons which are attracted to the positive end of the metal. The unit of electric current is the ampere (symbol A). This is defined as the current along two infinitely long thin wires 1 metre apart in a vacuum when a force of 2.0×10^{-7} newtons per metre acts on each wire due to their magnetic fields.

★ The charge passing a point in a circuit in a certain time interval is defined as the current × time taken. The unit of charge is the coulomb (symbol C) which is the charge passing a point in a circuit in 1 second when the current is 1 A.

Certain insulating materials become charged with electricity when rubbed with a dry cloth. Electricity as a word was coined from the Greek word for 'amber' by William Gilbert in the sixteenth century. Gilbert investigated the attractive force of amber and certain other materials when rubbed and concluded that such materials were charged with electricity as a result. More experiments showed that there are just two types of electric charge which we now refer to as 'positive' and 'negative'.

(i) Static electricity is due to gain or loss of electrons by an insulator or an insulated conductor. Certain insulating materials lose electrons easily and therefore can easily be charged positively by rubbing. Other insulating materials gain electrons easily and therefore can easily be charged negatively. Charged objects attract each other if they carry opposite types of charge and they repel if they carry the same type of charge.

(ii) Current electricity is due to the flow of charged particles (referred to as charge carriers in a solid). In metals and in intrinsic and n-type semiconductors, the charge carriers are electrons. In p-type semiconductors, the charge carriers are holes. In an ionic solution, the flow of charge is carried by positive and negative ions. Electrons move round a circuit from − to + because they are negatively charged particles. However, the direction of current in a circuit diagram is usually shown from + to − because this convention was established by Andre Ampere long before electrons were discovered.

see also...

Electric Circuits; Electrical Conduction

Circular Motion

An object in circular motion changes its direction of motion continuously. Its velocity is not constant because its direction of motion is not constant. The resultant force needed to change the direction of motion of an object in circular motion is called the centripetal force. Uniform circular motion means circular motion at constant speed.

'Centripetal' means 'towards the centre'. The centripetal force on an object in circular motion is perpendicular to its direction of motion as the direction of motion at any point is tangential. The centripetal force does no work on the object because the object does not move in the direction of the force as its path is a circle of constant radius.

★ The time taken, T, for such an object to move once round its circular path is the circumference/speed. Hence $T = 2\pi r/\upsilon$, where r is the circle radius.
★ The angular speed, ω, of an object in uniform circular motion = $2\pi/T$. This is the angle in radians that the radial line to the object moves through each second. The unit of ω is the radian per second (rad s^{-1}). Combining $T = 2\pi r/\upsilon$ and $\omega = 2\pi/T$ gives $\upsilon = \omega r$.
★ The centripetal acceleration of an object in uniform circular motion is always directed to the centre of the circle. The magnitude of the centripetal acceleration $a = \omega^2 r = \upsilon^2/r$. Hence the centripetal force $F = m\omega^2 r = m\upsilon^2/r$ for an object of mass m (because $F = ma$; see p.40).

Someone in a fairground vehicle or an aircraft that descends rapidly then ascends experiences an extra support force of $m\upsilon^2/r$ at the bottom of the descent, where r is the radius of curvature of the vehicle path, υ is the speed at the bottom and m is the person's mass. This extra support force is in addition to the person's weight, mg, and is sometimes referred to as the 'g-force'. For example, if υ^2/r is equal to 3 g, the person experiences a g-force equal to 3 times his or her weight.

see also...

Force and Motion; Satellite Motion

Colour 1 – The White Light Spectrum

The spectrum of white light consists of the colours of the rainbow and has a continuous spread of wavelengths from about 350 nm (violet) to about 650 nm (red). Colour vision is due to the effect of light on the three types of colour-sensitive cells, called cones, which are on the retina of the eye. Each type of cone has its peak sensitivity in a different part of the wavelength range, corresponding to blue, green and red light. These colours are the primary colours of light.

★ When you observe a white object in white light, photons of all wavelengths in the range from 350 to 650 nm are incident on the surface and are scattered by it. Some of these photons enter your eyes, causing all three types of cones to be stimulated which is interpreted by your brain as white light.

★ When you observe a particular colour of the rainbow, photons of a certain wavelength stimulate one or more of each of the types of cones. For example, yellow light consists of light photons of wavelength about 600 nm which stimulate red-sensitive cones and green-sensitive cones. The brain learns to interpret this pattern of stimulation as yellow light.

★ A secondary colour is seen if two spotlights of different primary colours overlap on a white screen. The secondary colours are yellow (= red + green), magenta (= red + blue) and cyan (= blue + green). For example, you observe yellow where red and green circles overlap. This is because your red-sensitive cones are stimulated by red photons and your green-sensitive cones are stimulated by green photons. Your blue-sensitive cones are not stimulated. You see yellow in both this case and the preceding example because your red and green cones are stimulated even though in the above example this is due to 600 nm photons only whereas in the case of the overlapping circles of green and red light colour, it is due to 650 nm (i.e. red) photons and 550 nm (i.e. green) photons.

see also...

Colour 2; Photon

Colour 2 – Colour Filters and Pigments

The observed effect of filters and surface pigments on light depends on the absorption of light photons by the molecules of the filter or the pigment.

★ When white light is passed through a colour filter, the filter molecules absorb photons of a certain range of wavelengths. When you observe the transmitted light, the three types of cones of the retina are stimulated according to the photons reaching the retina. This is interpreted by the brain as the colour of light passing through the filter. For example, a yellow filter absorbs photons of blue light so only red- and green-sensitive cones are stimulated.

★ When you observe a coloured surface in white light, pigment molecules on the surface absorb photons of certain wavelengths. The scattered light therefore does not contain these photons. For example, a yellow surface absorbs blue photons from white light and scatters all the other photons. Thus the red- and green-sensitive cones of the retina are stimulated so a yellow surface is seen.

★ When you observe a coloured surface in coloured light, the pigment molecules absorb photons of the same wavelengths as are absorbed from white light if these are present in the incident light. If so, the scattered light does not contain these photons. For example:

(i) if green light is directed at a red shirt, the shirt appears black because the pigment molecules of the red shirt can absorb photons of all colours except red.

(ii) if a yellow shirt is placed in a spotlight of white light and a cyan filter is placed between the shirt and the light source, the shirt appears green in colour. This is because the cyan filter removes red light from the white light and the shirt absorbs the blue photons from the filtered light. The remaining photons therefore stimulate the green cones of the retina more strongly than the red or blue cones of the retina.

see also...

Colour 1; Photon

Dark Matter

One of the biggest mysteries in science at the start of the twenty-first century is the whereabouts of most of the matter in the Universe. This hidden matter is known as dark matter or 'missing mass'. Dark matter is invisible matter hidden in galaxies or between galaxies but is known to be present as it slows rotating galaxies down. It constitutes at least 90% of the mass of the Universe yet it cannot be detected directly.

The presence of dark matter in vast quantities has been deduced by studying stars in the arms of spiral galaxies. By calculating the speeds of such stars from their measured Doppler shifts, astronomers have deduced that spiral galaxies rotate and that the stars in the arms of a spiral galaxy take the same length of time for each complete rotation, regardless of distance from the galactic centre. The total mass of the galaxy can be estimated from its rate of rotation. A star at the outer edge of a spiral galaxy keeps moving round the galactic centre because it experiences a force of gravitational attraction to the galactic centre in the same way as a planet moving round the Sun. However, the further a planet is from the Sun, the longer its orbit, unlike the stars in the arms of a spiral galaxy which mostly go round with the same time period, regardless of distance. To produce the result that the time period is independent of radius, it is necessary to assume the galaxy includes much more matter in its spiral arms than can be accounted for by the stars there. The problem of missing mass thus arises because if all the mass of a galaxy was in the stars of the galaxy, the galaxy should be much brighter than it actually is. The luminosity-to-mass output of a typical galaxy is less than a tenth that of a typical star. Since the light from a galaxy is due entirely to its stars, at least 90% of the mass of a typical galaxy must be outside its stars and therefore hidden.

Dark matter could be due to **neutrinos** which radiate from stars in in vast quantities.

see also...

Gravitational Fields 1 and 2;
Satellite Motion

Decay Processes

A decay process is a process where a quantity decreases at a decreasing rate. An exponential decay process is where the rate of decrease of the quantity is proportional to the quantity. Decay processes occur in capacitor discharge and in radioactive decay. The mathematics of decay processes can also be applied to physical processes such as the absorption of radiation by materials.

Any exponential decay process can be modelled numerically by starting with the condition for an exponential decay, namely the rate of decay of a quantity is proportional to the quantity itself. A numerical model of exponential decay can be displayed using a computer program or spreadsheet. For example, suppose the number N of radioactive nuclei of a given isotope decreases by 10% every hour and the initial number of radioactive nuclei present is 10 000.

The table below shows how N decreases every hour. The time taken for the number of particles to decay to 50% of the initial number is about 6.5 hours. Show for yourself that the number of particles will fall to about 50% of 5000 in another 6.5 hours. The time taken for the number of particles to decrease to 50% is called the half-life of the decay process.

The same idea applies to any process in which a quantity decays exponentially; the **half-life** $T_{1/2}$ of the process is the time taken for the quantity to reduce to 50%. For example, if the p.d. of a capacitor decreases by 10% every second when it discharges through a resistor, the capacitor would take about 6.5 seconds to discharge half its initial charge.

see also...

Radioactivity 1

Time from start/hours	0	1	2	3	4	5	6	7	8	9	10	11	12
Number of particles present	10 000	9000	8100	7290	6561	5905	5314	4783	4305	3874	3487	3138	2824
Decrease per hour	1000	900	810	729	656	591	531	478	431	387	349	314	282

Decibels

Sound consists of vibrations that travel through solids, liquids and gases in waves. Any vibrating surface in air creates sound waves that travel through the air away from the surface. The sound waves repeatedly compress then expand the air by tiny amounts. The frequency of the sound waves is the number of compressions per second at any place.

The loudness of a sound depends on frequency as well as intensity. Maximum sensitivity is at a frequency of about 3400 Hz. No sound can be detected beyond a frequency of about 18 000 Hz.

★ The intensity level, in decibels, of a sound of intensity $I = 10 \log (I/I_0)$, where $I_0 = 10^{-12}$ W m^{-2}, the least detectable intensity of sound at

Intensity in terms of I_0	1	10^2	10^4	10^6	10^8	10^{10}	10^{12}
Intensity level in decibels (dB)	0	20	40	60	80	100	120
Typical sound	None	Whisper	Rustle	Talk	Shout	Bellow	Engine roar

The **intensity** of a sound wave is the energy per second incident normally on unit area of a surface.

Sound waves make your eardrum vibrate. The larger the vibrations, the louder the sound you hear. The ear's response to sounds of different loudness at the same frequency is logarithmic, which means that equal increases of loudness are due to equal percentage increases of intensity. So, if the intensity is doubled, then doubled again then doubled again, the increase of loudness is the same each time the intensity is doubled.

1000 Hz for a person with normal hearing.

★ The loudness of a sound at a certain frequency is defined as the intensity level of a sound of equal loudness at a frequency of 1000 Hz. The unit of loudness is the phon. For example, a 100 dB sound at a frequency of 10 000 Hz has the same loudness as a 40 dB sound at 1000 Hz. The 100 dB sound therefore has a loudness of 40 phons.

see also...

Wavemotion 1 and 2

14

Diffraction

Diffraction is the spreading of waves after passing a gap or by the edge of an obstacle. Diffraction is important in optical devices such as microscopes and telescopes and in communications. Wavefronts from a point source spread out as expanding spheres. All the points on a wavefront vibrate in phase with each other. At a sufficiently large distance from the source of the waves, the wavefronts over a small enough region are effectively straight and are referred to as plane wavefronts. The propagation of a wavefront was explained in the seventeenth century by Christian Huyghens in his theory of wavelets; each point on a wavefront is considered as a secondary emitter of waves or wavelets in the direction in which the wave is advancing. The wavelets from the points along a wavefront create a new wavefront which creates more wavelets which create a new wavefront and so on. Huyghens used his theory to explain reflection and refraction.

The amount of diffraction of waves passing through a gap increases if the gap narrowed or if the wavelength is increased. When diffraction at a gap occurs, only part of each wavefront passes through the gap. Each wavefront emerging through the gap is shorter than it was. The wavelets from each restricted section spread out to recreate wavefronts that spread out. The greater the restriction of the wavefronts, the more significant the effect of wavelets from the ends of the restricted wavefronts so the greater the spreading. When waves pass an obstacle, the waves spread behind the obstacle if the wavelength is of the same order as the obstacle's width.

Diffraction of light passing through apertures and lenses in optical devices restricts the detail that can be seen in the optical image formed by the device. Adjacent features of the image overlap if too much diffraction occurs and therefore cannot be seen as distinct and separate features. By using wide lenses in optical devices, diffraction is reduced and image detail is better.

see also...

Optical Images 2; Wavemotion 1 and 2

Dynamics

Displacement is distance in a specified direction. Speed is rate of change of distance. Velocity is speed in a specified direction. The unit of speed and of velocity is the metre per second (m s⁻¹). The average speed of an object = distance/time taken. Acceleration is rate of change of velocity. Velocity changes occur due to change of direction as well as change of speed. The unit of acceleration is the metre per second per second (m s⁻²).

The dynamics equations for constant acceleration

For an object moving along a straight line at constant acceleration a, its speed changes at a constant rate. If its initial speed is u, then its speed at time t later, $v = \boldsymbol{u} + \boldsymbol{at}$.
Also, since its speed changes at a constant rate, its average speed v_{AVE} = ½ $(u + v)$. Hence the distance travelled, $\boldsymbol{s}$ = average speed × time taken = ½ $(\boldsymbol{u} + v)\boldsymbol{t}$.
Combining the two equations in bold gives $s = ut + \frac{1}{2}at^2$ (by eliminating v), and $v^2 = u^2 + 2as$ (by eliminating t).

Velocity time graphs (for motion in a straight line). A velocity time graph is a graph of velocity on the y-axis against time on the x-axis. The + y-axis represents one direction and the − y-axis represents the opposite direction.

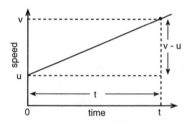

(i) The gradient of the line is equal to the acceleration. A negative gradient corresponds to a negative value of acceleration which is a deceleration.
(ii) The area between the line and the x-axis represents the displacement, taking areas below the line as the distance moved in one direction and areas above the line as distance moved in the opposite direction.

> ## see also...
> *Force and Motion*

Efficiency

The efficiency of a device = useful energy from the device/ energy supplied to the device = useful output power/input power. Efficiency can be expressed as a percentage by multiplying either ratio above by 100. A machine makes things move as a result of using either mechanical or electrical energy. The efficiency is the fraction of the supplied energy that is used usefully. A heat engine makes things move as a result of using thermal energy, usually from fuel. A transformer produces electrical power at a certain alternating voltage due to being supplied with electrical power at a different alternating voltage. Energy that is not used is wasted, usually as internal energy of the surroundings. This energy cannot be recovered in the sense that it cannot all be used to do more work. At a national level, about 20% of the nation's energy supplies from fuel is wasted due to inefficiencies in power stations, in distribution and in refineries.

A heat engine is any engine that operates between a high temperature 'reservoir' and a low temperature 'reservoir'. Heat from the high temperature reservoir is used to do work. Not all the heat from the high temperature reservoir can be converted to work as some heat must be supplied to the low temperature reservoir because the engine operates on the basis of a temperature difference. The efficiency of an engine = W/Q_1 where W is the work done by the engine when it is supplied with energy Q_1 from the high temperature reservoir. Since $W = Q_1 - Q_2$, where Q_2 is the energy transferred to the low temperature reservoir, then the efficiency of the engine = $(Q_1 - Q_2)/Q_1$. The efficiency must be less than 1 as Q_2 must be non-zero.

The most efficient heat engine is an idealised 'reversible engine'. By definition, the efficiency of a reversible engine $= \dfrac{T_1 - T_2}{T_1}$ where T_1 is the temperature of the high temperature reservoir and T_2 is the temperature of the low temperature reservoir.

see also...

Energy and Power; Entropy

Elasticity

Elasticity is that property of a material which allows it to regain its shape when it is released. Consider a length l of material of uniform cross-sectional area A. When under tension T, the material extends to a length $l + e$, where e is the extension of the material.

★ The tensile stress in the material
$$= \frac{\text{tension } T}{\text{area of cross-section } A}$$
The unit of tensile stress is the pascal (Pa) which is equal to 1 $N\,m^{-2}$.

★ The strain in the material
$$= \frac{\text{extension } e}{\text{initial length } l}$$
Strain is a ratio and therefore has no unit.

★ The Young modulus of elasticity E of a material
$$= \frac{\text{tensile stress}}{\text{strain}}$$
provided the limit of proportionality of the material is not exceeded. The unit of E is the pascal (Pa) which is equal to 1 Nm^{-2}.

The relationship between stress and strain for many materials shows certain common features:

1 Stress is proportional to strain, up to a limit known as the limit of proportionality. The stress/strain value is constant and equal to the Young modulus of the material. Up to the limit of proportionality, the tension is therefore proportional to the extension.

2 The elastic limit is the point beyond which a material does not regain its initial shape when the tension is removed. The atoms of the material are pulled out of place beyond the elastic limit so the material cannot regain its initial shape.

3 When a material is stretched beyond its elastic limit, it gives a little at the 'yield point' then regains its strength as the stress is increased.

4 As the tension is increased beyond the yield point, the material stretches and the stress increases and a neck forms. Further stretching causes the stress to concentrate at the neck until it breaks. The breaking stress is the breaking force/the initial area of cross-section.

see also...

Forces in Equilibrium; States of Matter

Electric Circuits

(i) **A series circuit** has components connected together so they always pass the same current. In a series circuit, charge flows along one path only through every component in sequence.

1 The current through components in series is the same for each component.

2 The potential difference across a series combination is equal to the sum of the potential differences across the individual components.

3 For two or more resistors R_1, R_2, R_3, etc. in series, their combined resistance $R = R_1 + R_2 + R_3 + \cdots$, etc.

All the electrons that enter a component each second leave it so the current leaving a component is the same as the current entering the component. The current passing through two or more components in series is the same because the same electrons pass through each component. Note that 'same' does not mean 'equal'. Two components in different branches of a circuit might pass equal currents but they do not pass the same current.

Components in series are all switched on or off together by a switch in series with the components. A fuse in a plug is always in the live wire in series with the appliance element or motor so that the appliance disconnects from the live wire if the fuse blows.

(ii) **In a parallel circuit**, charge flows from one point to another along parallel paths.

1 The potential difference across components in parallel is the same for each component.

2 The current through a parallel combination is equal to the sum of the currents through the individual components.

3 For two or more resistors R_1, R_2, R_3, etc. in parallel, their combined resistance R is given by the equation $1/R = 1/R_1 + 1/R_2 + 1/R_3 + \cdots$, etc.

Components in parallel can be switched on or off independently by a switch in series with each component. Appliances connected to a ring main circuit and light sockets connected to a lighting circuit are connected in parallel with each other so they can be switched on or off independently.

> ### see also...
> *Charge and Current; Potential Difference and Power; Kirchoff's Laws*

Electric Fields 1 – Electric Field Strength

An electric field is a region surrounding a charged object where a force acts on any other charged object in that region. The lines of force of an electric field are lines along which a small positively charged object would move if free to do so.

(i) The strength of an electric field, E, at a point in an electric field is defined as the force per unit charge acting on a small positively charged object at that point. The unit of electric field strength is the newton per coulomb ($N\,C^{-1}$) which is the same as the volt per metre ($V\,m^{-1}$). Hence the force F on a point charge q at a point in an electric field $= qE$, where E is the electric field strength at that point.

(ii) A uniform electric field exists between two oppositely charged parallel conducting plates at fixed separation. The lines of force are parallel to each other at right angles to the plates. Because the field is uniform, its strength is the same in magnitude and direction everywhere. The potential increases uniformly from the negative to the positive plate along a line of force. For a potential difference V_p between the plates, the work done to move a point charge q from one plate to the other $= qV_p$ so the force F on q = work done/distance $= qV_p/d$, where d is the distance between the plates. Therefore the electric field strength $E = F/q = V_p/d$.

(iii) A radial electric field surrounds a point charge. The field lines are directed outwards if the point charge is positive and inward if the point charge is negative. Consider a particle with a small charge q in the electric field created by a particle carrying a much larger charge Q. The two particles exert equal and opposite forces on each other, given by **Coulomb's Law** of force, $F = Qq/4\pi\varepsilon_0 r^2$, where ε_0 is the absolute permittivity of free space and r is the distance between the two particles, therefore the electric field strength at q's position due to Q is $E = F/q = Q/4\pi\varepsilon_0 r^2$.

Note that Coulomb's Law is an example of an inverse square law since F is inversely proportional to the square of the distance r.

see also...

Electric Fields 2; Inverse Square Laws

Electric Fields 2 – Dielectrics

A dielectric is an insulating substance that weakens the force between charged objects when the substance is present between the objects. Charged objects can be brought together or separated more easily when a dielectric is present between them. In an electric field, the molecules of a dielectric become polarised to produce a reverse polarisation field which effectively reduces the applied field strength. In a capacitor, this dielectric action enables more charge to be stored at the same p.d., thus increasing the capacitance of the capacitor. Water is very effective as a dielectric which is why ionic solids dissolve in it.

If a dielectric substance is placed between two oppositely charged parallel plates connected to a battery, the amount of charge stored on the plates increases because of the presence of the dielectric substance. This occurs because the dielectric substance in effect weakens the electric field between the plates, allowing the battery to push more charge on to the plates. The ratio of the charge stored on the plates of a parallel plate capacitor with the dielectric substance present to the charge stored without the dielectric substance present is known as the **relative permittivity** (also referred to as the dielectric constant) of the substance, ε_r.

The capacitance of a pair of parallel plates $C = Q/V$ where Q is the charge stored at a plate p.d. V. With a dielectric substance between the plates, the charge stored is increased by a factor ε_r for a fixed p.d. Therefore, the capacitance is increased by placing a dielectric substance between the plates. In a practical capacitor, the 'plates' consist of two strips of metal foil separated by a dielectric substance, all rolled up into a tube. The greater the relative permittivity of the dielectric substance, the greater the capacitance of the capacitor. The voltage applied to a capacitor containing a dielectric substance should never exceed the maximum working voltage (usually printed on the capacitor) otherwise the dielectric substance will break down and conduct.

see also...

Capacitance; Electric Fields 1

Electric Conduction

Conduction in metals, intrinsic semiconductors and n-type semiconductors is due to the presence of conduction electrons as the charge carriers. These electrons are free to move through the material because they are not held by any ion. The charge carriers in a p-type semiconductor are holes. In an ionic liquid, the charge carriers are positive and negative ions.

When a potential difference is applied across a metal or a semiconductor, the charge carriers are attracted towards the oppositely charged end of the material. As a result, the charge carriers gradually drift towards that end of the conductor.

For a material in the shape of a conductor of uniform cross-sectional area, its **conductivity** is defined as its length/(resistance × area of cross-section.)

The unit of conductivity is the siemens per metre ($S\,m^{-1}$).

The **resistivity** of a material = 1/conductivity. The unit of resistivity is the ohm metre ($\Omega\,m$). The conductivity of a material depends on the number of charge carriers per unit volume in the material. Increasing temperature causes the conductivity of a metal to decrease because the atoms in the metal vibrate more to impede the movement of conduction electrons through the metal by scattering them more. The conductivity of a semiconductor increases with an increase of temperature. This is because more electrons break free from the atoms in the semiconductor as its temperature is increased.

see also...

Charge and Current; Resistance

Classification	Conductivity / $S\,m^{-1}$	Resistivity / Ωm	Carrier density / m^3	Example
Conductor	about 10^6 or more	about 10^{-6} or less	about 10^{25} or more	any metal, graphite
Insulator	about 10^{-6} or less	about 10^6 or more	less than 10^{10}	polythene
Semiconductor at room temperature	about 10^3	about 10^{-3}	about 10^{20}	silicon, germanium

Electromagnetic Induction

Electromagnetic induction is the generation of a voltage due to changing magnetic flux linkage through a circuit.

★ Lenz's Law states that an induced electromotive force (e.m.f.) acts in a direction so as to oppose the change causing the induced e.m.f. This is a consequence of the fact that if the circuit is complete and has no other sources of e.m.f., the induced current creates a magnetic field which acts against the magnetic field that causes the induced e.m.f.
★ Faraday's Law of electromagnetic induction states that the induced e.m.f. in the coil is proportional to the rate of change of magnetic flux linkage through the coil.

The two laws of electromagnetic induction, Faraday's Law and Lenz's Law, apply to all situations where an e.m.f. is induced due to changing magnetic flux. The greater the rate of change of flux, the greater the induced e.m.f. If the circuit is complete, an induced current passes round the circuit. Electromagnetic induction can be due to the movement of a conductor in a magnetic field (e.g.

dynamo, alternator, microphone) or to changing the magnetic flux density through a coil (e.g. induction coil, inductors, transformer, tape recorder on playback). In a transformer, the alternating current through the primary coil produces a changing magnetic flux in the core and through the secondary coil so an alternating voltage is induced in the secondary coil. The ratio of the secondary voltage to the primary voltage is equal to the ratio of the number of secondary turns to primary turns.

When an electric motor is operating, a back e.m.f. is induced in the coil because the flux through the coil continually changes as the coil spins. This changing flux causes an induced e.m.f. or 'back e.m.f.' to act against the applied voltage. The back e.m.f. is proportional to the motor frequency. The current increases when the load on the motor is increased because the coil slows down and the back e.m.f. falls, allowing more current through the coil.

see also...

Alternating Current; Magnetic Fields 1

Electromagnetic Waves

Electromagnetic waves are electric and magnetic fields that propagate through space or a substance, vibrating in phase with each other as they progress. No substance is needed as the fields oscillate without the need for any substance.

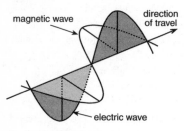

An electromagnetic wave

Electromagnetic waves were predicted by James Clerk Maxwell in 1862 who showed that an electromagnetic wave propagates through space at the speed of light. He concluded that light is therefore an electromagnetic wave and that electromagnetic waves should exist beyond the visible spectrum at either end. The full spectrum of electromagnetic waves is as follows, in order of increasing wavelength:

Gamma rays   X-rays
Ultraviolet light
Visible light
Infrared radiation
Microwaves
Radio waves

Maxwell's prediction of electro-magnetic waves beyond the infrared region were confirmed when in 1887 Heinrich Hertz discovered how to produce and detect radio waves. He showed that the speed of radio waves is the same as that of light.

Electromagnetic waves in the visible, infrared and radio parts of the spectrum are used extensively as carrier waves in communications. The higher the frequency, the greater the information that can be carried by the carrier waves which is why optical fibres can transmit much more information than ordinary copper wires can.

Electromagnetic waves in the X-ray and gamma region of the spectrum are used to form images of body organs and bones.

see also...

Colour 1; Photon; X-rays 1 and 2

Electron

The electron is a fundamental particle and is within every atom. The electron carries a fixed negative charge and has a well-defined mass and spin. It is one of six fundamental particles known as leptons. The other charged leptons, the muon and the tauon, carry the same charge as the electron and are only detected as a result of high energy collisons.

The charge of the electron, e, is 1.60×10^{-19} C. All other charged particles except quarks carry a charge which is a whole number $\times e$. The specific charge of the electron, e/m, is its charge divided by its mass. The value of e/m is 1.76×10^{11} C kg^{-1}. Its spin, also referred to as its intrinsic angular momentum, is equal to $^1/_2\, h/2\pi$. Hence it is referred to as a spin $^1/_2$ particle.

The electron was discovered in 1897 by J.J. Thomson who showed that cathode rays produced in a discharge tube consist of identical negatively charged particles. He showed that the specific charge of the electron is much larger than that of any other charged particle and proved that the specific charge of the electron is 1.76×10^{11} C kg^{-1}. Thomson made his discoveries about the electron as a result of a series of experiments involving the deflection of a beam of cathode rays in electric and magnetic fields. Thomson deduced that electrons are negatively charged particles within every atom. He was not able to measure the charge or the mass of the electron so he was unable to conclude from his research on cathode rays that the electron is many times lighter than a hydrogen atom.

The charge on the electron, e, was measured in 1915 by Robert Millikan who invented a method of measuring the charge on individual charged oil droplets. Millikan discovered that the charge on an oil droplet was always a whole number $\times 1.6 \times 10^{-19}$ C. He concluded that charge is quantised in units of 1.6×10^{-19} C and that the quantum of charge, 1.6×10^{-19} C, is the charge carried by a single electron. The mass of the electron could then be calculated from its charge divided by its specific charge, thus giving a rest mass of 9.1×10^{-31} kg.

see also...

Electron Beams 1 and 2; Particle Interactions

Electron Beams 1 – Thermionic Emission

An electron beam is produced in a vacuum tube by the process of thermionic emission from a filament wire. In this process, the wire is heated by passing an electric current through it. Electrons in the metal wire gain sufficient kinetic energy to leave the metal and are attracted to a nearby positively charged metal plate which has a small hole in it to allow some electrons through. These electrons are then focused into a beam by passing through 'focusing' electrodes.

The kinetic energy and hence the speed of an electron in an electron beam depends on the anode potential V_A as the work done on each electron by the anode gives the electron its kinetic energy. Since the work done = eV_A, the kinetic energy of an electron in the beam is therefore equal to eV_A. Provided the speed υ of the electron is much less than the speed of light, its kinetic energy = $\frac{1}{2}m\upsilon^2$, therefore $\frac{1}{2}m\upsilon^2 = eV_A$.

An electron beam is deflected using electric and magnetic fields. From the above equation, it follows that all the electrons in the same beam have the same kinetic energy and speed and are therefore deflected by electric and magnetic fields equally. In practice, the electrons in the beam have a small range of speeds due to their relatively small initial kinetic energies in the filament wire.

★ In a TV tube or a VDU tube, magnetic deflecting coils are used to force the beam to scan across a fluorescent screen, successive scans being horizontal and displaced downwards slightly. In this way, an image is formed on the screen as the signal variations are used to control the brightness of the beam.

★ In an oscilloscope tube, electro-static deflecting plates are used to make the beam scan repeatedly along the same line, slowly in one direction then much more rapidly on return. A voltage waveform is displayed on the screen as a result of applying the voltage across a pair of parallel plates which the beam passes between.

see also...

Charge and Current; Electric fields 1 and 2; Electron; Magnetic Fields 1 and 2

Electron Beams 2 – Trajectories in Fields

In a uniform electric field of strength E, an electron in a beam experiences a constant force $F = eE$ in the opposite direction to the field direction. The path of the beam is therefore a parabola, the same as the path of a projectile since a projectile experiences a constant force (i.e. its weight) in one direction only. A uniform electric field of strength $E = V/d$ is produced by applying a constant potential difference V between two parallel metal plates at a spacing d. In an oscilloscope, the deflection of the electron beam is proportional to the potential difference between the deflecting plates.

angles to the field direction and to the direction of motion of the electron. In the absence of any other fields, the electron therefore moves round on a circular path of radius $r = m\upsilon/Be$. This is because the force causes centripetal acceleration hence $Be\upsilon = m\upsilon^2/r$. The magnetic field does no work on the electrons in the beam since the force is at right angles to the direction of motion of the beam. In a TV tube or a VDU tube, the magnetic field used to force the beam to scan across a fluorescent screen is produced by passing a suitably changing current through a set of deflecting coils.

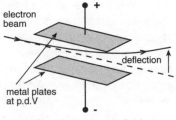

In an electric field

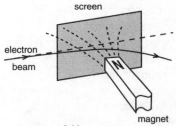

In a magnetic field

In a uniform magnetic field of flux density B, an electron moving at speed υ at right angles to the lines of the magnetic field experiences a force $F = Be\upsilon$. The force direction is at right

see also...

Circular Motion; Electric Fields 1 and 2; Electron Beams 1; Magnetic Fields 1 and 2

Energy and Power

★ Work is done by a force on an object when the object is moved by the force. The amount of work done by a force is defined as 'force × distance moved in the direction of the force'. The unit of work is the joule, equal to the work done when a force of 1 newton moves its point of application by 1 metre in the direction of the force.

★ Energy is the capacity to do work. The unit of energy is the joule. If work is done on an object, the energy of the object is increased. If the object does work, its energy decreases.

★ The Principle of Conservation of Energy states that the total amount of energy in an isolated system remains constant.

★ Power is the rate at which energy is transferred. The unit of power is the watt (W), equal to 1 J s^{-1}.

The **kinetic energy** of a moving object is the energy it possesses due to its motion. For an object of mass m moving at a speed υ, its kinetic energy $E_k = \frac{1}{2}\, m\upsilon^2$, provided its speed is much less than the speed of light. The formula $E_k = m\,c^2 - m_0\,c^2$ from the theory of special relativity must be used for speeds that are not insignificant compared with the speed of light.

The **potential energy** of an object is the energy it possesses due to its position relative to one or more other objects. The unit of potential energy is the joule. If an object is moved above the Earth so its height changes, its potential energy changes because of the force of gravity on the object. Since the force of gravity on an object of mass m due to the Earth is equal to mg, the energy transferred when an object of mass m is raised a height h above the Earth = force × distance moved along the line of action of the force = mgh, where g is the gravitational field strength. Hence the change of potential energy of the object = mgh. This formula does not apply if the height h is significant compared with the radius of the Earth as g is significantly less further from the Earth. In this case, the change of potential energy is worked out using a formula derived from Newton's Law of Gravitation.

see also...

Force and Motion; Gravitational Fields 1

Energy from the Nucleus

The nucleus of every atom except the hydrogen atom consists of protons and neutrons held together as a result of the strong nuclear force, which acts equally between protons and neutrons, has a range of no more than about 2 or 3 $\times 10^{-15}$ m and is much stronger than the electrostatic force of repulsion between the protons.

The binding energy of a nucleus is defined as the energy that must be supplied to a nucleus to separate it into its constituent protons and neutrons. This energy is needed to overcome the strong nuclear force holding the protons and neutrons together. Energy supplied to any object increases its mass in accordance with Einstein's equation $E = mc^2$. Hence the mass of any nucleus is less than the mass of the separated protons and neutrons. This mass difference is known as the mass defect Δm of the nucleus. The binding energy E_B of any isotope $^A_Z X$ can thus be calculated using the formula $E_B = c^2 \Delta m = c^2 (Zm_p + (A-Z)m_N - M)$, where m_p, m_N and M represent the rest masses of the proton, the neutron and the nucleus respectively. Also, Z is the number of protons in the nucleus

and $(A - Z)$ is the number of neutrons. (See p.4.)

The graph of binding energy per nucleon, E_B /A, against nucleon number for all the known nuclei below shows that the most stable nuclei occur at about $A = 50$ where the binding energy per nucleon is greatest.

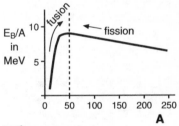

Binding energy curve

★Fusion of two light nuclei to form a nucleus no larger than $A = 50$ releases energy because the fused nucleus is more tightly bound than the light nuclei used.
★Fission of a heavy unstable nucleus into two fragments releases energy because the fragment nuclei are more tightly bound than the heavy nucleus.

see also...

Fission; Fusion; Radioactivity 1

Energy Level Models

Models of the atom that explain energy levels are based on the wave-like nature of electrons. A hydrogen atom consists of an electron trapped in the electrostatic field of a proton. The electron is trapped like an object in a well, and can only exist in the 'potential well' at certain energy levels.

(i) **A rectangular potential well** is the simplest model of a hydrogen atom. If the width of the well is L, the electron can be considered as a standing wave across the well. Hence its de Broglie wavelength λ is such that $n\lambda/2 = L$, where n is a whole number. Hence its momentum $m\upsilon = h/\lambda = nh/2L$ so its kinetic energy $E_k = \frac{1}{2}m\upsilon^2 = (nh/2L)^2/2m = E_1\,n^2$, where $E_1 = h^2/8mL^2$. The total energy of an electron in the well $= E_k - eV_0$, where V_0 is the depth of the well. Thus the deepest electron energy level in the well is $E_1 - eV_0$, the next one is at $4E_1 - eV_0$, etc. This simple model produces energy levels but because the levels are not in agreement with experimental measurements, the model is an over-simplification.

(ii) The exact pattern of the energy levels of the hydrogen atom was deduced from the discovery that the photon frequencies emitted by hydrogen atoms fit a formula of the type $hf = E_1\,(1/n^2 - 1/m^2)$, where n and m are whole numbers. The energy levels occur at values of $-E_1/n^2$. These values were explained by Erwin Schrödinger who devised a general equation which could be applied to any charged particle in any potential well. The energy level formula above follows from the $1/r$ nature of the electrostatic potential surrounding the nucleus. Schrödinger's equation also gives the shape of the allowed 'probability shells' of the electrons in an atom which are the most probable locations of electrons in the atom. In addition, the equation provides a partial explanation of the number of electrons allowed in each shell. This explanation is completed by the Pauli Exclusion Principle.

see also...

Energy Levels in Atoms; Exclusion Principle; Wave Particle Duality

Energy Levels in Atoms

An energy level is a possible energy value of a system of two or more particles. Energy levels occur in any confined system where the quantum nature of the particles is important. This is the case wherever the de Broglie wavelength of the particles is significant compared with the distance between the particles.

Energy levels in atoms were discovered in electron collision experiments using gas-filled electron tubes. Electrons are emitted from the hot filament in the tube and attracted to the anode. When the anode potential is increased, electrons reach the anode through a metal grid thus causing a current to the anode. The anode current falls then rises at certain values of the anode potential, referred to as excitation potentials, as the potential is increased.

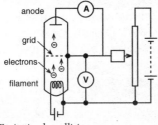

Excitation by collision

Each dip in the current happens when electrons from the filament with just sufficient kinetic energy collide with gas atoms and cause electrons in the atoms to gain energy and move to a higher energy level. The electrons from the filament are stopped by the collisions and attracted onto the grid instead of moving onto the anode. Thus the anode current is at a minimum at an excitation potential. The gas atom is excited to a higher energy level. The energy gained by an excited gas atom is equal to the kinetic energy of a beam electron since each beam electron gives up all its kinetic energy in the process. Hence the energy levels occur at energies eV above the ground state, where V is any excitation potential. The energy needed to ionise an atom (i.e. to remove an electron from an atom completely) is equal to eV_0, where V_0 is the potential needed to ionise the atom. Thus the lowest energy state of an atom, referred to as its ground state, is at energy eV_0 below the ionisation level.

see also...

Energy Level Models; Wave Particle Duality

Energy Resources

Energy supplies are obtained from fossil fuels such as coal, oil and gas, and from non-fossil fuels such as biomass and uranium. Coal formed from decayed vegetation and oil and gas formed from dead marine life forms, both

★ World fuel use is about 400×10^{18} J per year. This is presently met from energy sources as shown below. The list also shows how many years present fuel reserves will last at the 1995 rate of use.

Oil	Coal	Natural gas	Nuclear power	Hydroelectricity
40%	27%	22%	8%	3%
50 years	300 years	70 years	60 years*	indefinitely
		[*for thermal reactors (3000 years for fast breeder reactors)]		

compressed by successive layers of rock over millions of years.

Renewable energy resources such as hydroelectricity, tidal power and geothermal power make a significant contribution to energy supplies in

★ UK fuel use and reserves are shown below. The total energy used by all sources in the United Kingdom is about 3.5% of the world fuel use. The list also shows the lifetime of present UK fuel reserves at the 1995 rate of use.

Oil	Coal	Natural gas	Nuclear power	Hydroelectricity
59%	13%	21%	7%	<0.5%
<50 years	1300 years*	40 years	imported	indefinitely
		[*based on 1983 estimates if coalfields closed since 1983 are re-opened]		

some parts of the world. A renewable energy resource can supply energy via processes that do not use fuel and do not waste materials. Solar power, wave power and wind power are renewable energy resources likely to become increasingly important as oil and gas reserves dwindle if nuclear power is phased out.

Renewable energy resources could supply more power as fuel reserves dwindle. In Britain, wind turbines and hydroelectric power stations now contribute to the electricity grid.

see also...

Efficiency; Energy and Power

Entropy

ntropy is a measure of the disorder of a system. The disorder of a system can be measured in terms of the number of ways in which the particles and the energy of the system can be arranged. The greater the number of arrangements, the more disordered a system is.

The entropy S of any system is defined as $S = k \ln W$, where W is the number of ways the particles of a system can be arranged and k is the Boltzmann constant. (See p.1.) This definition gives the result that heat transfer Q to or from a system at absolute temperature T causes an entropy change of the system $\Delta S = Q/T$. The unit of entropy is J K^{-1} (or J K^{-1} mol^{-1} for 1 mole of a substance).

The Second Law of Thermodynamics states that it is impossible for heat transfer from a high temperature source to produce an equal amount of work. Some of the energy is wasted as heat transfer to a low temperature 'sink' which is necessary to keep working. Thus energy tends to spread out irreversibly and become less useful when work is done so the entropy of the system increases.

A reversible change is one that can be reversed to bring a system back to its original state. For example, a pendulum bob released from a non-equilibrium position will swing across and return exactly to its original position, provided there is no air resistance.

Most changes are not reversible as they cause energy to spread out irretrievably. This spreading occurs because it is the most likely outcome of all the possible rearrangements that the change could cause. For example, imagine a box partitioned into two halves and a small hole in the partition. Suppose initially there are just four molecules moving about in one half of the box. A distribution of two molecules in each half will be the most likely outcome after a sufficiently long time. There are 16 ($= 2^4$) ways of distributing the molecules between the two halves. The most probable outcome is with two on either side because there are six ways for this distribution to be arranged which is more than for any other distribution.

see also...

Efficiency; Ideal Gases

Evolution of Stars

The evolution of a star is the sequence of stages it passes through from its formation as a protostar to the end of its life as a light-emitting object. A star forms from clouds of dust and hydrogen gas as a result of the inward pull of gravity by matter in the clouds on other matter in the clouds. As the protostar becomes denser, gravitational energy is converted to thermal energy and the temperature of the protostar increases until it is hot enough for nuclear fusion to commence. High energy radiation released at this stage heats the protostar even more and it becomes stable, as a main sequence star.

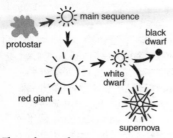

The evolution of a star

A star stays on the main sequence for much of its life, emitting radiation as a result of hydrogen nuclei being fused into helium nuclei in its core.

Radiation released in this process exerts pressure on the layers of the star outside the core. The gravitational pull on each layer of a star is balanced by the outward force due to radiation pressure. When all the hydrogen nuclei in its core have been used, the core collapses and the outer layers swell out and cool to become a red giant. At this stage, the helium nuclei in the core fuse to form nuclei as heavy as iron. When this process ends, the star collapses and heats up to form a white dwarf if its mass is less than 1.4 solar masses. If the white dwarf is one of two stars in a binary system, it may draw matter onto itself from its binary companion. If this happens, it flares up to become a nova.

If the mass of a star exceeds 1.4 solar masses, known as the Chandresakhar limit, the star collapses completely and then explodes as a **supernova**. Such a massive explosion causes the formation of heavy nuclei as the shock wave of the explosion makes light nuclei fuse together.

see also...

Fusion; Nuclear Model of the Atom

Exclusion Principle

An electron in an atom has a certain amount of energy and it occupies a shell which is its most probable position according to its energy. Each shell can hold no more than a certain number of electrons, referred to as its occupancy number. The innermost shell can hold no more than two electrons and the next shell no more than eight electrons. The arrangement of elements in the Periodic Table is accounted for by these occupancy numbers. For example, the neon atom in its ground state has two of its 10 electrons in the innermost shell and the other eight in the second innermost shell. Neon is an unreactive element because the electrons of its atoms are in full shells.

The occupancy levels of the electrons in the atom were explained in 1925 by Pauli who realised that the state of each electron in the atom is specified by four quantum numbers which no other electron in the atom can possess. This is known as Pauli's Exclusion Principle.

★ The energy E of an electron in the nth shell is given by $E = E_1/n^2$, where E_1 is the energy of an electron in the $n = 1$ shell. The shell number n is referred to as the principal quantum number.

★ The angular momentum of an electron in a shell is quantised in multiples of a basic amount where the multiple, the angular momentum quantum number, ℓ, is any whole number from zero to $n-1$.

★ The magnetic quantum number, m_l, due to the orbiting electron acting as a tiny magnet, takes values from $+\ell$ to $-\ell$.

★ The spin of an electron, s, is its intrinsic angular momentum. Pauli realised that an electron in an atom exists in two possible spin energy states, one with its spin parallel to the spin of the nucleus and the other lower energy state with its spin in the opposite direction.

The occupancy numbers work out as follows for the first two shells: 1st shell: $n = 1$, $\ell = m_l = 0$ holds two electrons (as there are two spin states), 2nd shell: $n = 2$, $\ell = m_l = 0$ holds two electrons and $\ell = 1$, $m_l = \pm 1$ or 0 holds six electrons, therefore eight electrons in total.

see also...

Electron; Energy Levels in Atoms; Types of Bonds

Fission

Fission is the splitting of a nucleus into two approximately equal fragments. Uranium 235 and plutonium 239 are the only known isotopes that release neutrons when fissioned. Uranium 235 is the only naturally occuring fissile isotope.

A large nucleus is like an oscillating liquid drop. If an oscillating nucleus is struck by a neutron, the nucleus forms two parts which repel each other electrostatically. The two fragments release high energy neutrons which can go on to produce further fission. If more than one neutron per fission goes on to produce further fission, an uncontrollable chain reaction occurs. In a nuclear reactor, a steady chain reaction is established in which fission occurs at a precise rate of 1 further fission per fission neutron released. The other fission neutrons are absorbed by nuclei in the control rods and elsewhere without fission or they escape from the reactor. The energy released per kilogram of fissile material is about a million times more than the energy released per kilogram of oil when oil burns. Uranium 235 and uranium 238 are the two naturally occuring isotopes of uranium with abundances of less than 1% and about 99%, respectively.

★ In a thermal nuclear reactor, the fuel rods consist of enriched uranium which contains of the order of 2 to 3% uranium 235. The fission neutrons are too fast to produce further fission so a moderator surrounding the fuel rods is used to slow the fission neutrons down so they can produce further fission. The fission neutrons lose kinetic energy when they collide with the moderator's nuclei.

★ In a fast breeder reactor, neutrons released from plutonium 239 produce further fission without the need for a moderator. Plutonium 239 is an artifical isotope created when uranium 238 absorbs a neutron and decays. Thus a fast breeder reactor can produce its own fuel in the form of plutonium 239 from a blanket of uranium 238 surrounding the core. On a large scale, this process would extend the availability of nuclear fuel by a factor of 50, in effect extending the lifetime of the world's reserves of uranium by many centuries.

see also...

Nuclear Model of the Atom;
Nuclear Power

Flow Processes

An electric current is a flow of charge due to a potential difference, heat transfer by conduction is a flow of energy due to a temperature difference and fluid flow is a flow of liquid or gas due to a pressure difference. General flow characteristics thus include some sort of difference which drives particles or energy from one region to another.

The flow rate is the rate of flow or the amount of charge or energy or matter passing per second in a certain direction.

★ In an electric circuit, the current through a wire depends on the potential difference across the wire and the resistance of the wire in accordance with the equation current = potential difference / resistance.

★ In a conductor of heat of uniform cross-sectional area, the heat flow depends on the temperature difference and the thermal resistance of the conductor in accordance with the equation heat transfer per second = temperature difference / thermal resistance. The thermal resistance is equivalent to the electrical resistance which is the resistivity of the material × the conductor length / area of cross-section.

★ In a pipe through which fluid flows, the flow rate depends on the pressure difference between the ends of the pipe and the resistance to flow of the fluid. The equivalent equation for the flow rate to the previous two equations is mass per second flowing through the pipe = pressure difference / pipe resistance. The pipe resistance depends on the viscosity of the fluid as well as on pipe dimensions. If the fluid viscosity is negligible, there is no resistance to flow and no pressure difference is needed to keep the fluid passing along the pipe. The pipe's internal surface drags on a viscous fluid moving through it. As a result, the resistance is raised much more by using a narrower pipe than happens if an electrical or thermal conductor is replaced by a narrower conductor.

see also...

Fluids 2; Heat Transfer; Resistance

Fluids 1 – Fluids at Rest

A fluid is any substance that can flow. For any fluid at rest, the pressure at any point in it acts equally in all directions and increases with depth. The pressure at the base of a column of liquid exceeds the pressure at the top of the column by an amount equal to $h\rho g$, where h is the height of the column, ρ is the density of the liquid and g is the strength of gravity on the Earth. (See p.44.) To prove the formula, consider the volume of the column which is equal to its height h × its area of cross-section A. Hence the mass of liquid in the column $m =$ volume × liquid density = $hA\rho$. Thus the weight of the liquid in the column = $mg = hA\rho g$ so the pressure on the base of the column due to its weight = weight of liquid / area of cross-section = $hA\rho g / A = h\rho g$.

An object in a fluid experiences an upthrust because the pressure of the fluid on its base pushing up is greater than the pressure of the fluid on the top of the object pushing down. In a fluid of density ρ, a vertical cylinder of cross-sectional area A and height h experiences a pressure difference equal to $h\rho g$ between its base and its top. Hence the upthrust on the cylinder is equal to pressure difference × area of cross-section which is equal to $h\rho gA$. Since hA is the volume of the cylinder, then $h\rho gA$ is the weight of the fluid displaced by the object. Thus the upthrust is equal to the weight of the fluid displaced by the object, as discovered by Archimedes and known as **Archimedes' Principle**.

1 The weight of fluid displaced when an object is fully immersed must be more than the weight of the object if the object is not to sink. The density of the object must therefore be less than the density of the fluid for flotation. If the object's density exceeds the density of the fluid, the object will sink.

2 A ship or a boat floats lower in the water when it is loaded. The vessel displaces more water when it is loaded so the upthrust increases until it is equal to the increased weight of the vessel. The vessel will sink if it is loaded to the point of total immersion as the upthrust cannot exceed the weight of water displaced at total immersion.

see also...

Fluids 2; Pressure

Fluids 2 – Fluids in Motion

Viscosity is the property of a fluid that determines how difficult it is to make it flow. For example, oil does not pour as quickly as water because the viscosity of oil is higher than that of water. Some fluids such as paint become less viscous when stirred. Some fluids such as wallpaper paste become more viscous when stirred.

Fluid flow is said to be streamlined if a marker dye released into the flow at a point follows a definite line and does not break into turbulence. The speed of the fluid, its density, its viscosity and the presence of boundaries and edges determines whether flow is streamlined or turbulent. Turbulence occurs if the forces tending to change the momentum of the fluid exceed the viscous forces. This is determined by the **Reynolds number** R which is defined as $\rho \upsilon D / \eta$, where ρ is the density of the fluid, υ is its speed, η is its viscosity and D is the width of the object in the fluid. If R is less than about 2000, the flow is streamlined.

Non-viscous flow is where the viscosity of a fluid is negligible. For streamlined flow of a non-viscous fluid, low pressure occurs where the flow is fast and high pressure occurs where the flow is slow. This is known as the **Bernouilli Principle** and it follows from conservation of energy as any change of the fluid's kinetic and potential energy is due to work done by the pressure forces in the fluid. For flow in which changes of potential energy are negligible, the pressure is higher where the speed is lower, and the pressure is lower where the speed is higher. The lift force on an aircraft wing is because the wing is shaped so the air flow is faster over the top than underneath. Hence the air pressure underneath is greater than on the top so an upward force acts on the wing. Stalling occurs if the angle between the wing and the direction of motion exceeds a certain value that depends on the wing speed. Turbulence then occurs and the airflow speed over the wing drops so the pressure above the wing is no longer less than the pressure below it.

see also...

Fluids 1; Force and Motion; Pressure

Force and Motion

★ The **momentum** of an object is defined as its mass × its velocity. The unit of momentum is the kilogram metre per second (kg m s⁻¹). Momentum is a vector quantity.

★ Newton's 1st Law of motion states that an object remains at rest or moves with constant velocity unless acted on by a resultant force. Newton's 1st Law defines what a force is, namely any physical effect that is capable of changing the motion of an object. If an object is at rest or in uniform motion, either no force acts on it or forces do act on it and the resultant force is zero.

★ Newton's 2nd Law of motion states that the rate of change of momentum of an object is proportional to the resultant force on the object.

Consider an object of constant mass m, acted on by a constant force F so its speed increases from u to v in time t. Because the force is proportional to the change of momentum / time taken, then $F = k (mv - mu)/t$ where k is a constant. Since acceleration $a = (v - u)/t$, then $F = kma$. By defining the unit of force, the newton (N), as the amount of force that gives a 1 kg

mass an acceleration of 1 m s⁻², then $k = 1$ and Newton's 2nd Law for constant mass may be written as $F = ma$ provided the force is in newtons, the mass is in kg and the acceleration is in m s⁻².

★ Newton's 3rd Law states that two objects exert equal and opposite forces on each other when they interact. This is sometimes stated as 'To every action, there is an equal and opposite reaction.'

The Principle of Conservation of Momentum states that the total momentum of a system of bodies is always the same provided no external forces act on it. In a collision between two objects or an explosion where two objects fly apart, the momentum of each object changes. Because the two objects exert equal and opposite forces on each other for equal times, one object gains momentum at the expense of the other object so the total change of momentum is zero. The total momentum is therefore conserved.

see also...

Dynamics; Forces in Equilibrium

Forces in Equilibrium

An object at rest acted on by several forces is in static equilibrium because the forces on the object balance each other out and have no overall turning effect on the object.

(i) The forces on an object balance out if the force vectors joined end to end form a closed polygon. The combined effect of two or more forces acting on an object can be determined by means of a vector diagram in which the force vectors join end-on; the resultant force vector is straight from the tail of the first vector to the tip of the last one. The object is in equilibrium if the resultant is zero (i.e. the tip of the last one is at the tail of the first one) which means that the force vectors must form a closed polygon for equilibrium.

(ii) The overall turning effect is zero if the individual turning effects of the forces about the same point balance out. The turning effect of a force about a point is called the **moment** of the force and is defined as the product of the force and the perpendicular distance from the line of action of the force to the point. The rule that the overall turning effect must be zero for an object in equilibrium is known as the **Principle of Moments** and is usually expressed as the statement that the sum of the clockwise moments about a point must equal the sum of the anticlockwise moments. The conditions for static equilibrium of an object acted on by several forces are therefore:

1 the resultant force is zero, which corresponds to the force vectors forming a closed polygon,

2 the Principle of Moments applies about any point.

Static equilibrium can be either neutral, stable or unstable, according to whether or not an object displaced slightly from its equilibrium position stays in its new position (neutral) or returns to its equilibrium position (stable) or moves away (unstable). An object such as a high-sided vehicle will topple over if it is tilted too much. This occurs if tilting causes its **centre of gravity**, the point where all its weight may be considered to act, to move beyond its wheel base.

see also...

Force and Motion; Vectors

Fusion

Fusion is the process of joining together light nuclei to form a heavier nucleus. Energy is released in this process provided the nucleus formed contains no more than about 50 neutrons and protons. To make two nuclei fuse together, they must approach each other to within a distance of 2–3×10^{-15} m which is the range of the strong nuclear force. The initial kinetic energies of the two nuclei to be fused must be of the order of MeV to overcome the electrostatic repulsion between the nuclei and thus enable them to approach each other to within 2–3×10^{-15} m. This is achieved inside a star as a result of its extremely high core temperature which is maintained by the energy released due to fusion of hydrogen nuclei (protons) to form helium and other nuclei. The energy released per helium nucleus formed is about 7 MeV per nucleon which is considerably more than the energy released in fission.

Fusion can be created in a fusion reactor by using magnetic fields to contain a plasma of ionised hydrogen and passing a very large current of the order of 10^6 A through it. This heats it sufficiently to cause fusion of the protons to form helium nuclei as well as heavier nuclei, in stages as below.

1 $p + p \rightarrow {}^2_1H + {}^0_{+1}\beta + 0.4$ MeV (in the plasma).

2 ${}^2_1H + {}^3_1H_1 \rightarrow {}^4_2He + {}^1_0n + 17.6$ MeV (in the plasma).

3 ${}^6_3Li + {}^1_0n \rightarrow {}^4_2He + {}^3_1H + 4.8$ MeV (in a blanket of lithium surrounding the core).

The tritium (${}^3_1H$) formed in the lithium blanket is removed from the blanket and fed into the plasma. Neutrons released in the plasma are absorbed by the lithium nuclei to form tritium nuclei and helium nuclei. Thus the process releases a total of 22.8 MeV for every four protons and neutrons forced to form a helium nucleus. The raw materials are hydrogen and lithium which form helium. In theory, the energy released by fusion should maintain the high temperature of the plasma. However at present, fusion in such a reactor cannot be sustained to produce more power than is supplied to it.

see also...

Nuclear Model of the Atom

General Relativity

The general principle of relativity is that the laws of physics are the same for all observers. In 1916, Albert Einstein published his General Theory of Relativity in which he proved mathematically the general principle of relativity. In this theory, he showed that absolute space and absolute time are meaningless and replaced them with the concept of space time by proving that space and time are interdependent. In essence, his theory is that matter distorts space time and space time makes matter move. He showed that the distortion of space time is in proportion to the distribution of matter and energy. He showed that Newton's Law of Gravitation follows from his theory if gravity is weak enough.

Einstein had started a revolution in physics in 1905 when he published the Special Theory of Relativity. He was in his mid-twenties at the time and was employed as a full-time patent officer in Berne. He became a full-time university lecturer in 1909 and moved to Berlin in 1913 as director of a research institution specially set up for him. In 1916, Einstein published the General Theory of Relativity in which he predicted the existence of black holes and the bending of light by gravity. Einstein's theory was successfully tested by Sir Arthur Eddington who photographed stars near the Sun during the total eclipse of 1919. Eddington discovered that the position of the stars in line with the edge of the Sun's disc were slightly displaced as Einstein had predicted. The successful test of Einstein's General Theory meant that concepts such as absolute space and absolute time were no longer correct. Space and time are interlinked and affected by gravity. Einstein became an international celebrity when a conference of leading scientists discussing his theory was reported the next day in the *Times*. The General Theory of Relativity has had important consequences for astronomy and cosmology, including the discovery of black holes, gravitational lensing and the Big Bang theory of the origin of the Universe.

see also...

Big Bang; Black Hole; Gravitational Fields 1

Gravitational Fields 1 – Strength of Gravity

A gravitational field is the region surrounding an object where a force due to the mass of the object acts on any other object in that region. The lines of force of a gravitational field are lines along which a small mass would move if free to do so.

The strength of the gravitational field, g, at a point in a gravitational field is defined as the force per unit mass acting on a small mass at that point. The unit of gravitational field strength is the newton per kilogram (N kg^{-1}). The force F on a point mass m at a point in a gravitational field $= mg$, where g is the gravitational field strength at that point. This is therefore the weight of an object of mass m.

Newton's Law of Gravitation states that the force of gravitational attraction, F, between any two masses m_1 and m_2 is proportional to $m_1 \times m_2$ and inversely proportional to the separation r between the two centres of mass.

Hence $F = G\, m_1 \times m_2/r^2$, where G is the constant of proportionality and is referred to as the Universal Constant of Gravitation. The value of G has been accurately measured and its accepted value is 6.67×10^{-11} N m^2 kg^{-2}.

The force of gravitational attraction on a small mass m near a large spherical planet of mass M is therefore $F = G\,M\,m/r^2$, where r is the distance from m to the centre of M. Therefore the gravitational field strength $g = F/m = G\,M/r^2$ at distance r from the centre of the planet. Also, the surface gravitational field strength $g_s = G\,M/R^2$, where R is the radius of the planet. The surface gravitational field strength at the Earth's surface varies with latitude from 9.81 N kg^{-1} at the poles to 9.78 N kg^{-1} at the equator. This is because of the spinning motion of the Earth and the fact that the equatorial radius is slightly more than the polar radius.

see also...

Force and Motion; Projectiles

Gravitational Fields 2 – Escape Speed

A rocket needs to attain a speed of about 11 kilometres per second to escape from the Earth's surface and reach the Moon or beyond. This minimum speed is known as the escape speed from the Earth's surface. If the engines of a rocket are not powerful enough, the rocket does not attain escape speed so it falls back to Earth.

It can be shown that the energy needed for a rocket of mass m to escape from the surface of a planet of mass M is equal to GMm / R, where R is the planet's radius. If this energy is to come from the rocket's kinetic energy after using all its fuel, then it must attain kinetic energy at least equal to GMm / R to escape. Hence its escape speed v_{esc} must be such that the minimum kinetic energy for escape, $\frac{1}{2} m v_{esc}^2$, is equal to GMm / R. Thus the escape speed from the surface of a planet = $\sqrt{(2GM / R)}$ which is equal to $\sqrt{(2g_s R)}$ since g_s, the surface gravitatiional field strength, is equal to GM / R^2. At the surface of the Earth, $g = 9.80$ N kg^{-1} and $R = 6370$ km approximately. Hence the escape speed is $\sqrt{(2 \times 9.80 \times 6370 \times 1000)}$ = 11 200 metres per second. At the surface of the Moon, $g = 1.62$ N kg^{-1} and $R = 1740$ km, hence the escape speed from the Moon = 2380 metres per second. The much smaller escape speed from the Moon is the reason why the Apollo astronauts who landed on the Moon were able to return from the lunar surface to their lunar orbiters in modules much smaller than the giant Saturn rockets which were needed to escape from Earth.

The Earth has an atmosphere whereas the Moon does not. This is because gas molecules in the Earth's atmosphere move at speeds far below the escape speed of 11.2 km s^{-1} and hence are unable to escape from the pull of the Earth's gravity. Gas molecules released on the Moon would have a similar range of speeds as molecules in the Earth's atmosphere as the temperature range on the Moon is similar to that on the Earth. These gas molecules would escape from the Moon because the escape speed is much lower from the lunar surface.

see also...

Energy and Power; Gravitational Fields 1

Heat Transfer

Heat is energy transferred due to a difference of temperature. The three methods of heat transfer are conduction, convection and radiation.

Conduction occurs in solids, liquids and gases. Metals are the best conductors of heat because of the presence of conduction electrons which gain kinetic energy when the metal is heated. These electrons transfer energy from the hot parts of a metal to the cooler parts as a result of colliding with electrons and atoms in the cooler regions. Heat transfer in non-metals, liquids and gases is caused by the motion of atoms in the hot regions spreading to atoms in the other regions.

The thermal conductivity of a material is the heat transfer per second per unit area of cross-section in the material for a temperature gradient of 1 K per metre. For a uniform insulated conductor of cross-sectional area A and length L which has a temperature difference ΔT across its ends, the heat transfer per second from one end to the other, $Q/t = k A \Delta T / L$.

Convection is heat transfer due to hot fluid circulating and displacing cooler fluid. In general, fluid rises where it is heated because its density is less than that of the cooler fluid. For example, hot air rises from an electric heater in a room and forces the air to circulate.

Thermal radiation is electromagnetic radiation emitted by a surface due to its temperature. The hotter an object is, the more thermal radiation it emits. A surface that is a good absorber of thermal radiation is also a good emitter. A matt black surface is the most effective absorber and a shiny silvered surface is the least effective.

The spectrum of the thermal radiation from a surface at temperature T is continuous and peaks at a certain wavelength λ_p in accordance with Wien's Law, $\lambda_p T = 0.0029$ K m. Stefan's Law states that the total energy emitted per second per unit area, $W / A = \sigma \varepsilon T^4$, where σ is the Stefan constant and ε is the emissivity of the surface.

see also...

States of Matter; Temperature

Hubble's Law

dwin Hubble used the 2.5 m reflector telescope on Mount Wilson in California to estimate the distances to two dozen galaxies with known red shifts within six million light years of the Milky Way galaxy. His results, published in 1929, showed that the red shift increased with distance. By plotting the results on a graph of red shift against distance, it was clear that the red shift and hence the speed of recession is in proportion to the distance. This relationship is known as Hubble's Law. The constant of proportionality in the relationship is known as the Hubble constant, H.

light years. Subsequent measurements have reduced the Hubble constant to its present-day value of about 20 km s^{-1} per million light years.

Hubble's Law is an experimental law valid for a limited range of measurements. It is now accepted that Hubble's Law follows because the Universe is expanding from a primordial explosion between 10 000 and 15 000 billions of years ago. This explosion, known as the Big Bang, was the origin of space and time. The value of H is used to estimate the age of the Universe. In simple terms, the speed of the furthest galaxies

$$\text{Speed of recession} = \text{the Hubble constant} \times \text{distance}$$
$$v \qquad\qquad H \qquad\qquad d$$

Further measurements of more galaxies were made by Milton Humason. By 1935, Hubble and Humason had measured more than 140 galaxies out to distances of more than 1000 million light years moving away at speeds of over 40 000 kms^{-1}. The results confirmed Hubble's findings of 1929 that the red shift increased with distance. Hubble and Humason estimated the Hubble constant to be 160 kms^{-1} per million

cannot exceed the speed of light, c, which is 300 000 km s^{-1} so the distance to such galaxies cannot exceed c/H. Taking account of gravity gives $2c/3H$ which works out at about 12 000 million light years.

see also...

Big Bang; Red Shift

Ideal Gases

The experimental gas laws are:

★ Boyle's Law, which states that the product of pressure × volume is constant for a fixed mass of gas at constant temperature,
★ Charles' Law, which states that the increase of volume of a fixed mass of gas at constant pressure is proportional to the increase of its temperature,
★ The pressure law which states that the increase of pressure of a fixed mass of gas at constant volume is proportional to its increase of temperature.

An ideal gas is a gas under such conditions that it obeys Boyle's Law. The experimental gas laws are combined in a single equation, the ideal gas equation. This relates the number of moles n, the pressure p, the volume V and the absolute temperature T of an ideal gas to one another in the form $pV = nRT$, where R is the molar gas constant. The value of R is 8.31 J mol^{-1} K^{-1}. The ideal gas equation can be derived by making certain assumptions about the gas:

1 A gas consists of point molecules of equal mass.

2 The molecules collide elastically with each other and with the container walls.
3 The molecules are in continual random motion.
4 The molecules do not attract each other.
5 The duration of impact between a molecule and the container walls is much shorter than the time between successive collisions with the walls.

Using Newton's laws and the rules of statistics applied to random events, the assumptions lead to the kinetic theory equation $pV = 1/3 \, Nmc_{rms}^2$, where N is the number of molecules present, m is the mass of a molecule and c_{rms} is the root mean square speed of the molecules of the gas. This is equal to the square root of the mean value of the squares of the molecular speeds. By making the assumption that the mean kinetic energy of a gas molecule $1/2 mc_{rms}^2 = 3/2 \, kT$ where $k = R/N_A$ (N_A = the Avogadro constant), the kinetic theory equation becomes $pV = nRT$.

see also...

Pressure; States of Matter

Interference

The Principle of Superposition states that when two or more waves overlap, the resultant displacement is equal to the sum of the individual displacements at that point and at that instant. Interference is the superposition of two or more waves that are coherent and therefore maintain a constant phase relationship.

Interference is produced if waves from two coherent sources overlap or if waves from a source are divided then reunited.

1 Interference using sound waves requires two loudspeakers connected together to an oscillator. The loudspeakers produce sound waves of the same frequency with a constant phase difference. The loudspeakers are therefore coherent sources of sound waves. Anyone moving about in the overlap area ought to be able to detect points of reinforcement and of cancellation, corresponding to positive and negative interference, respectively.

★ At a point of reinforcement, the sound is loud as a result of the wave peaks from one speaker arriving at the same time as wave peaks from the other speaker or wave troughs from one speaker arriving at the same time as wave troughs from the other speaker.

★ At a point of cancellation, the sound is quiet as a result of the wavepeaks from one speaker arriving at the same time as wavetroughs from the other speaker. Interference of light cannot be observed using two separate sources as light photons are emitted at random from separate sources.

2 Interference can be produced by division of the wavefronts from a source of constant frequency. Two narrow closely spaced slits allow part of each wavefront through so each slit acts as a source of waves that diffract from the slit. Interference occurs in the overlap area. The two slits are coherent emitters because there is a constant time difference between the two slits emitting waves from the same wavefront. This method, referred to as double slit interference, can be used to demonstrate interference of sound, of microwaves and of light.

see also...

Diffraction

Inverse Square Laws

An inverse square law is where a physical quantity such as radiation intensity or field strength at a certain position is inversely proportional to the square of the distance from a fixed point. For example, the intensity of radiation from a light bulb that emits equally in all directions falls to one quarter if the distance from the bulb is doubled and to one ninth if the distance is trebled. The explanation of the inverse square law is that the physical quantity spreads out evenly in all directions from a point without absorption. Thus the reading of a detector decreases with distance from the source because the detector collects less and less the further away it is from the source. Imagine a sphere with the source at the centre. At a distance r from the source, the quantity spreads out over the entire area of the sphere which is $4\pi r^2$. The amount passing through a unit area of the sphere is therefore inversely proportional to the sphere area and thus inversely proportional to r^2.

The following quantities obey the inverse square law:

1 Intensity of radiation from a point source, $I = k/r^2$, where k is a constant and r is the distance from the source, provided the radiation is not absorbed by a substance surrounding the source. For a source that emits radiation energy at a rate W, $k = W / 4\pi$, since all the radiation energy emitted per second passes through a sphere of surface area $4\pi r^2$ at a distance r. Therefore, since the intensity is defined as radiation energy per second per unit area, then $I = W/4\pi r^2$.

2 Electric field strength E at distance r from a point charge Q in a vacuum, $E = Q/4\pi\varepsilon_{0}r^2$. The lines of force radiate from Q. At a distance r, the effect of the charge Q has to cover a surface of area $4\pi r^2$ so $Q / 4\pi r^2$ is proportional to the field strength at a distance r from Q.

3 The gravitational field strength, g, at a distance r from the centre of a sphere of mass M is $g = GM/r^2$. The lines of force outside M are directed towards the centre of M. The r^2 factor represents the surface area of a sphere of radius r which the field has to cover.

see also...

Electric Fields 1; Gravitational Fields 1

Ionisation

onisation is the process of formation of ions. An ion is a charged atom or molecule. Positive ions are formed from certain types of atoms which easily lose outer-shell electrons. Negative ions are formed from those types of atoms which easily accept electrons. A free radical is a group of atoms which carries a charge, usually negative. The ionisation energy of an atom is the energy needed to ionise the atom. This is sometimes expressed in electron volts (eV) where 1 eV = 1.6×10^{-19} J.

Ionisation in a gas can be caused by a strong electric field or by heating the gas to sufficiently high temperature or by collisions between gas atoms or by high energy radiation. A lightning conductor causes air to conduct because a strong electric field is created at the tip of the conductor when a charged cloud is overhead. The field is strongest where the tip is sharpest so air molecules that enter this region are ionised which enables charge to leak between the tip and the thundercloud.

(i) In a gas discharge tube, a strong electric field is created in the gas at low pressure by means of a sufficiently large potential difference applied between two electrodes in the gas. As a result, electrons are pulled off the gas atoms which therefore become positive ions.

(ii) Heating a gas to several thousand degrees causes ionisation by collision of gas atoms at high speed. Two atoms lose some kinetic energy in such a collision and one or more of their outer electrons become detached. Light is emitted when the electrons and gas ions recombine. Inside a star, matter exists in an ionised state as the kinetic energies of the particles are too great for recombination to take place.

(iii) High energy radiation such as alpha and beta particles or gamma photons causes ionisation in solids, liquids and gases. High energy particles and photons from the Sun ionise gas atoms in the Earth's upper atmosphere. These ions form a conducting layer known as the Appleton layer which is responsible for the reflection of radio waves from ground transmitters back to the surface at frequencies below about 30 MHz.

see also...

Energy Levels in Atoms; Optical Spectra 1 and 2

Kirchoff's Laws

(i) Kirchhoff's 1st Law states that the total current entering a junction is equal to the total current leaving the junction. This is a statement of conservation of charge as it means that the total charge flowing into a junction in a given time is equal to the total charge leaving the junction in the same time.

Using the convention that currents leaving a junction are the opposite sign to currents entering the junction, the 1st Law may be expressed as the following equation:
$$i_1 + i_2 + i_3 + \cdots, \text{ etc.} = 0,$$
where i_1, i_2, i_3, etc., represent the currents in the branches connected to the junction.

(ii) Kirchhoff's 2nd Law states that the sum of the e.m.f.s round any complete loop in a circuit is equal to the sum of the potential drops round the loop. This is a statement of conservation of energy since an e.m.f. is where energy is supplied to charges (i.e. a source) and a potential drop is where charges release energy (i.e. a sink). The sum of the e.m.f.s is therefore the total electrical energy produced in the loop per unit charge and the sum of the potential drops is the total electrical energy used in the loop per unit charge.

For a loop of a circuit containing e.m.f.s E_1, E_2, E_3, etc. and resistances R_1, R_2, R_3, etc., the 2nd Law may be written as the following equation:
$$E_1 + E_2 + E_3 + \cdots, \text{ etc.} =$$
$$i_1 R_1 + i_2 R_2 + i_3 R_3 + \cdots, \text{ etc.},$$
where i_1, i_2, i_3, etc., represent the currents through the resistances R_1, R_2, R_3, etc.

Notes:

1 E.m.f.s and currents in the opposite direction to the direction round the loop are given negative values.

2 The 2nd Law is particularly useful to analyse circuits with more than one loop. In general, for a circuit with n unknown currents, n loops need to be considered one by one, each loop giving a new equation. In this way, the n linear equations formed can be solved for the n unknown currents.

see also...

Electric Circuits; Resistance

Liquid Crystals

L iquid crystals consist of molecules that can be arranged in an ordered pattern without being permanently and rigidly linked together. Such substances can flow even though their molecules may be in an ordered pattern as in a crystal. The essential condition of a crystalline material is that the atoms or molecules form an ordered pattern. However, liquid crystal does not have a characteristic shape but its molecules do form an ordered arrangement. Liquid crystal display systems are widely used because they require much less power than other display systems. Portable computers, hand-held electronic games and calculators are just three examples of liquid crystal display systems.

A liquid crystal display system is made up of a matrix of 'pixels' which are individual liquid crystal cells. Each cell contains a small amount of liquid crystal sandwiched between two transparent conducting surfaces parallel to each other. The surfaces in contact with the liquid are scored with fine parallel lines, perpendicular to each other. As a result, the alignment of molecules twists across the cell through 90°. This has the effect of rotating the plane of polarisation of polarised light that passes through the cell normal to the surfaces. The cell is fixed between two polaroids on top of a mirror. No light can pass through the cell when there is no potential difference between the surfaces so the cell appears dark from above. However, when a potential difference is applied between the two surfaces, the liquid crystal molecules align parallel to the field and do not affect the plane of polarisation of light passing through normally, so enabling light to pass through the polaroids. As a result, the cell no longer appears dark because light passing through it is reflected from the mirror. Each pixel appears dark or bright according to the potential difference applied to it. Liquid crystal displays are relatively slow to respond to changing potential differences in comparison with conventional display systems. This is because the molecules do not respond as rapidly as electron beams do in conventional display systems.

see also...

Polarisation; States of Matter

Magnetic Fields 1 – Magnetic Flux Density

A magnetic field is a force field created by a magnet or a current-carrying wire which acts on other magnets or current-carrying wires or on moving charged particles.

(i) A line of force in a magnetic field is the line along which a hypothetical free magnetic North pole would move if free to do so. A magnetic compass needle or any pivoted bar magnet aligns itself along a north–south plane in the Earth's magnetic field; the end pointing north is referred to as the 'north-seeking' pole and the other end is referred to as the south-seeking pole.

(ii) The strength of a magnetic field or magnetic flux density, B, is defined as the force per unit current per unit length acting on a current-carrying conductor placed perpendicular to the lines of a uniform magnetic field. The unit of B is the tesla (T), equal to 1 $N A^{-1} m^{-1}$. The direction of the force is perpendicular to the conductor and to the field.

The force F on a current-carrying conductor of length ℓ in a uniform magnetic field is given by the equation $F = Bl\ell \sin \theta$, where θ is the angle between the conductor and the lines of force of the field and B is the magnetic flux density of the field.

A charged particle moving across a magnetic field experiences a force at right angles to its direction of motion and to the lines of force. The force is given by the equation $F = Bq\upsilon \sin \theta$ where υ is the particle's speed, q is its charge and θ is the angle between the direction of motion of the charge and the field.

Magnetic flux: The magnetic flux ϕ through a surface of area A which is perpendicular to the lines of a uniform magnetic field is defined as BA, where B is the magnetic flux density. The magnetic flux linkage Φ through a coil of n turns and area A in a uniform magnetic field is defined as BAn, where B is the component of the magnetic field at right angles to the plane of the coil. The unit of magnetic flux is the weber (Wb), equal to 1 tesla per square metre $(T m^2)$.

see also...

Electromagnetic Induction; Mass Spectrometer; Particle Accelerators

Magnetic Fields 2 – Magnetic Materials

Ferromagnetic materials such as iron and steel can be magnetised permanently. Iron is easier to magnetise and demagnetise than steel which is why the core of an electromagnet is made of iron whereas a permanent magnet is made of steel.

If an unmagnetised bar of ferro-magnetic material is placed in a current-carrying coil, the material is magnetised and produces a magnetic field much stronger than the field due to the empty coil. The relative permeability of the material μ_r is defined as B / B_0, where B and B_0 are the magnetic flux densities created in a very long solenoid carrying a certain current with and without the presence of the material, respectively.

The relative permeability of a material is not a constant and depends on the strength of the magnetising field. This can be seen from a graph of B on the y-axis against I on the x-axis for a given material in a solenoid. The magnetic flux density increases non-linearly as the current is increased from zero to reach a constant level referred to as magnetic saturation, then decreases

to a non-zero value as the current is reduced from zero. The field is reduced to zero at a certain value of current in the reverse direction, known as the coercive current.

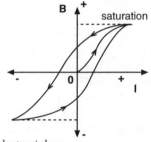

A hysteresis loop

Because the magnetic flux density lags behind the current, the B versus I curve is said to be a hysteresis loop. The relative permeability, μ_r, is proportional to B / I and can be up to about 2000 for iron. The area of the loop is a measure of the work done per cycle to magnetise and demagnetise the material. Iron therefore has a smaller loop area. Steel is much more difficult to demagnetise than iron so it has a high coercive current.

see also...
Magnetic Fields 1

Mass Spectrometer

n a mass spectrometer, ions are created from samples to be tested, usually by bombarding the sample with a beam of electrons. The ions are attracted to an oppositely charged electrode which has a hole in it to allow some of the ions to pass through and form a beam. A velocity selector is used to select ions in the beam moving at a certain speed. The ion beam is deflected by a magnetic field. Different ions are deflected by different amounts so the amount of deflection is measured and used to identify the ions and to measure the mass of each ion very precisely.

★ In the velocity selector, the beam of ions enters a uniform magnetic field at right angles to the beam and to a uniform electric field. The magnetic force on each particle, $Bq\upsilon$, is cancelled out by the electric force, qE, if the speed of the particles is such that $Bq\upsilon = qE$, where q is the charge of the particle. Because the ions are produced with a range of kinetic energies, only the ions in the beam at speed $\upsilon = E / B$ pass through undeflected. Thus the velocity selector is used in the mass spectrometer to select ions moving at the same speed.

★ The beam of ions is then directed into a uniform magnetic field at right angles to the beam. Each ion is forced on a curved path by the magnetic field. The centripetal force ($= m\upsilon^2/r$) on each ion is provided by the magnetic force ($= Bq\upsilon$) so the radius of curvature r of the curved path depends on the mass of the ions in the beam in accordance with the equation $r = m\upsilon/Bq$. Because all the ions in the beam travel at the same speed and experience the same magnetic force, the mass of an ion determines its path. Thus the beam is spread out by the magnetic field into different paths according to the masses of the ions present.

In a modern mass spectrometer, an electronic detector linked to a computer is used to measure the deflection of the ions and to calculate the mass of each type of ion present. The number of ions per second of each type is also measured from the detector current so the percentage abundance of each type of ion can be determined.

see also...

Circular Motion; Electron Beams 1 and 2

Moles and Mass

★ The Avogadro constant (symbol L or N_A) is defined as the number of atoms present in exactly 0.012 kg of $^{12}_{6}C$ (carbon 12). Carbon 12 is chosen because it can be separated easily from other carbon isotopes. This number has been measured accurately and is equal to 6.02×10^{23}.

★ One mole of a substance is the amount of substance when there are L particles present. Thus n moles of substance consisting of identical particles contains $n L$ such particles. The molar mass of a substance is the mass of one mole of that substance.

★ 1 atomic mass unit (u) is defined as $^1/_{12}$ th of the mass of a carbon 12 atom which by definition is therefore equal to 2.0×10^{-26} kg ($= 0.012$ kg / L). Hence 1 u $= ^1/_{12} \times 0.012$ kg / $L = 1.66 \times 10^{-27}$ kg. Note that the mass of a proton $= 1.007\ 28$ u, the mass of a neutron $= 1.008\ 66$ u and the mass of an electron $= 0.000\ 55$ u.

Because the mass of a proton and of a neutron is approximately equal to 1 u, the mass number of an isotope is therefore approximately equal to the mass in grams of one mole of the atoms of that isotope. For example, a nucleus of $^{238}_{92}U$ (uranium 238) consists of 238 neutrons and protons and therefore has a mass of approximately 238 u. Hence the mass of 1 mole of uranium 238 atoms is approximately 238 g or 0.238 kg.

★ The relative atomic mass of an atom or the relative molecular mass of a molecule is the mass of the atom or molecule in atomic mass units. Thus the molar mass of an element or compound is equal to its relative atomic or molecular mass in grams.

The number of atoms or molecules in mass m of an element or compound of molar mass M is equal to the number of moles (m / M) multiplied by the number of particles per mole, L. This type of calculation is used in radioactivity calculations where the number of atoms in a radioactive isotope has to be determined.

see also...

Atoms and Molecules

Nuclear Model of the Atom

Every atom contains a nucleus composed of protons and neutrons which are held together by the strong nuclear force acting between them. An isotope $^A_Z X$ contains Z protons and $A - Z$ neutrons.

Lord Rutherford proved the nuclear model of the atom by using alpha particles to probe the atom. He discovered that α particles in a narrow beam directed at a thin metal foil mostly passed through it. He measured the number of particles deflected per second through different angles and discovered that a small number were deflected through angles in excess of 90°. To explain these results, he assumed every atom contains a very small positively charged nucleus where most of its mass is located and that the nucleus repels an α particle because they carry the same type of charge. By applying Coulomb's Law, Rutherford showed that the number of particles per second deflected through an angle θ is proportional to $1/\sin^4(\theta/2)$, in agreement with his experimental findings. He estimated the diameter of the nucleus at about 10^{-5} times that of an atom and he recognised that the nucleus of the lightest atom, the hydrogen atom, is a single particle which became known as the proton. He also showed that the atomic number Z of an element is the number of protons in the nucleus of each atom of the element.

The neutron was predicted by Lord Rutherford on the grounds that the mass number of a nucleus is always more than its proton number so there must be neutral particles in the nucleus as well as protons. James Chadwick discovered the neutron as a result of bombarding beryllium foil with high energy alpha particles. He discovered that radiation emitted from the foil produced tracks in a cloud chamber due to collisions between particles of the radiation and the nuclei of nitrogen atoms. From the tracks, he was able to prove that the radiation consisted of uncharged particles of mass about the same as the proton.

see also...

Fission; Radioactivity 1–4

Nuclear Power

Nuclear power is produced as a result of the fission of uranium 235. The fuel rods in a thermal nuclear power station contain enriched uranium which is uranium 238 and 2–3% uranium 235. Each fission event causes a uranium 235 nucleus to split into two fragment nuclei, releasing two or three neutrons with individual kinetic energies of the order of MeV. The fragments recoil with kinetic energies of the order of 100 MeV or more, and they transfer this kinetic energy to neighbouring atoms. The fission neutrons cannot produce further fission unless they are slowed down.

The fuel rods are designed so that these fast neutrons leave the fuel rod and enter the surrounding substance or moderator. As a result of elastic collisions with the moderator nuclei, the fission neutrons lose kinetic energy until they have the same mean kinetic energy as a moderator nucleus. The fission neutrons move at random through the moderator and sufficient fission neutrons re-enter the fuel rods to cause further fission. In this way, a chain reaction continues inside the reactor core, maintained at a steady rate by control rods in the core which absorb excess neutrons to ensure exactly 1 neutron per fission goes on to produce a further fission event.

The core is enclosed in a sealed steel vessel through which a coolant fluid is pumped to remove the energy from the moderator. The hot coolant is passed through a heat exchanger to raise steam which is used to drive turbines and so generate electricity. The neutrons that escape from the core are either absorbed by the steel vessel or by the thick concrete walls in which the vessel is situated. The fuel rods are highly radioactive after being in the reactor vessel. This is because the fission fragments are neutron-rich and therefore beta emitters. In addition, the uranium 238 nuclei absorb neutrons without fission and become highly unstable, forming a series of radioactive isotopes including plutonium 239 which emit both gamma and beta radiation. The spent fuel rods are allowed to cool for up to a year in cooling ponds after removal from the reactor before being reprocessed to recover plutonium and unused uranium.

see also...

Fission; Radioactivity

Optical Images 1 – Using Mirrors and Lenses

When you look in a mirror, you see an optical image of yourself. It is a virtual image because it is where light scattered from each part of your face appears to come from after being reflected by the mirror. The image formed by a plane (i.e. flat) mirror is the same distance behind the mirror as the object is in front because the angle of reflection of any light ray reflecting off a mirror is the same as its angle of incidence, as shown below.

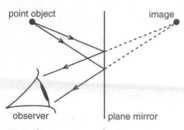

point object   image

observer   plane mirror

Image formation in a plane mirror

When you use a camera to take a photograph, light from the object being photographed is focused onto the film by the camera lens. The image on the film is a real image because it is formed as a result of the lens refracting light from the object onto the film. Refraction is the bending of a light ray when it crosses the boundary between two transparent substances. The lens is positioned so that all the light scattered from each point of the object into the lens is refracted to a single corresponding point on the film.

★ If the lens is not at the correct distance from the film, the image will be out of focus on the film because the light from each point of the object is not refracted to the same point on the film.

★ When a camera is used to take a picture of a distant object after taking a picture of a nearby object, the lens must be moved towards the film to refocus the image onto the film.

The **focal length**, f, of a convex lens is the distance from the lens to the image of a distant object. For an object at a distance u from the lens, the distance v from the lens to the image formed is given by the lens formula $1/u + 1/v = 1/f$. A positive value of v corresponds to a real image; a negative value to a virtual image.

> ## see also...
> *Optical Images 2*

Optical Images 2 – Image Quality

When a lens forms a real image of an object, light scattered by each point of the surface of the object is focused by the lens to form a tiny part of the image.

(i) The focal length, f, of a lens is the distance from the lens to where the image of a distant object is formed. The focusing power of a lens is defined as 1 / the focal length of the lens in metres. The unit of focusing power is the dioptre.

(ii) The magnification of a lens is the ratio of the size of the image to the size of the object. This depends on the distance from the object to the lens, and on the focal length of the lens. The image is smaller than the object if the distance from the object to the lens is greater than $2f$.

(iii) The amount of detail that can be seen in an image is a measure of the resolving power of the optical device used to form the image. Diffraction of light takes place as the light passes through the aperture of the device. A lens or curved mirror focuses light from a point of the object to a tiny spot not to a point image. Two nearby points of an object therefore form two nearby spots in the image. If diffraction is excessive because the aperture is too narrow, the nearby spots overlap and merge as a single a spot. The two points of the object are said to be 'unresolved' as they cannot be distinguised. By making the aperture sufficiently wide, diffraction is reduced so the two points are seen separately in the image. Thus the amount of detail visible in an image is improved by making the lens aperture wider, hence a 10 cm width telescope provides more detail than a narrower one. Ground-based telescopes wider than about 10 cm do not give better resolution because the Earth's atmosphere causes tiny amounts of refraction which smudges the images. The focal length of a lens therefore determines how much an image is magnified and the width of the lens determines how much detail can be seen. A short-focus lens gives more magnification than a long-focus lens but if the lens width is unchanged no more detail is seen because the resolving power of the lens is unaffected.

see also...

Diffraction; Optical Images 1

Optical Spectra 1 – Continuous Spectra

A spectrum is the spread of energies of the particles in a particle beam or of the photons in a beam of electromagnetic radiation. A photon is a wavepacket from less than 400 nanometres (nm) for violet light to more than 650 nm for red light. This narrow range of wavelengths is a small part of the electromagnetic spectrum.

Colour	violet	blue	green	yellow	orange	red
Wavelength / nm	<400	450	500	550	600	650
(approximate mid-range values)						

of electromagnetic radiation of a certain wavelength. Each photon has a definite amount of energy that depends on its wavelength. The spectrum of sunlight seen as a rainbow occurs because sunlight contains a continuous spread of wavelengths, and hence of photon energies. Each colour of light in the spectrum is due to photons of a certain energy. Refraction of sunlight by raindrops causes light of each colour to refract by a different amount forming a visible spectrum.

The spectrum of sunlight is an example of a continuous spectrum as the colour changes across the spectrum from violet to red without a break. By using a prism, sunlight and light from a filament light bulb can be split into a continuous spectrum of colours. The wavelength of light varies with colour,

A photon of light is emitted when an electron in a substance moves to a lower energy level. The energy of the photon is equal to the loss of energy of the electron. A filament light bulb or the Sun produces a continuous spectrum because the electrons in the glowing parts possess a continuous spread of energies producing a continuous range of photon energies. Any spectrum in which there is a continuous range of energies is described as continuous. For example, an X-ray tube produces photons with a continuous range of energies corresponding to wavelengths from about 0.001 nm or less to about 1 nm.

see also...

Electromagnetic Waves; Optical Spectra 2; Photon; X-rays 1 and 2

Optical Spectra 2 – Line Spectra

An optical line emission spectrum is a light spectrum in which characteristic coloured lines are seen at well-defined wavelengths only. Each coloured line is due to photons of a certain energy emitted by the light source. An atom emits a photon as a result of an electron in the atom transferring to a lower energy level. Light which forms a line spectrum is from a light source such as a vapour lamp or a discharge tube. The light-emitting atoms contain electrons at well-defined energy levels. The energy of a photon $E = hf = hc/\lambda$, where f is the frequency of the light, c is the speed of light and λ is its wavelength. If an electron transfers from energy level E_1 to a lower energy level E_2, the emitted photon has energy $hf = E_1 - E_2$. Because the energy levels of each type of atom are characteristic of that atom, the photon energies and therefore the pattern of wavelengths can be used to identify the element which the atom belongs to.

Absorption spectra are also produced by passing light through colour filters, coloured liquids and gels, and through coloured transparent solids. These materials absorb light over a range of wavelengths so the transmitted light is less intense over a range of wavelengths. An optical line absorption spectrum consists of narrow dark lines seen against a continuous spectrum. This type of spectrum may be seen in the spectrum of sunlight as a result of photons of particular wavelengths from the Sun's photosphere being absorbed by atoms in the gases surrounding the Sun. Such an atom is subjected to bombardment by photons of all possible wavelengths from the photosphere. Photons of certain energies only are absorbed by electrons in the atoms. Each such photon causes an electron to move from the inner shell to the outer shell of an atom. Light passing through the gas is less intense at certain wavelengths which therefore appear as dark lines against the continuous spectrum of sunlight. Line absorption spectra may also be produced in the laboratory by passing white light through a suitable gas or vapour before observing it using a slit and prism.

see also...

Energy Levels in Atoms; Optical Spectra 1; Photon

Particle Accelerators

An accelerator is a linear or circular device used to supply kinetic energy to charged particles. Each particle is accelerated by the electric field between pairs of electrodes. The high energy particles are then made to collide with other particles or anti-particles. The fragments of such collisions at very high energies produce new particles and anti-particles.

★ **The synchrotron** consists of an evacuated tube in the form of a ring which is fixed between the poles of a large number of electromagnets round the ring. Pairs of electrodes at several positions along the ring are used to accelerate charged particles as they pass. The electromagnets provide a magnetic field which keeps the charged particles on a circular path of fixed radius. The strength of the magnetic field is increased in synchronisation with the increase of particle mass so the radius of rotation is kept constant.

★ **The linear accelerator** consists of a long series of electrodes connected alternately to a source of alternating p.d. The electrodes are hollow coaxial cylinders in a long evacuated tube. Charged particles released at one end of the tube are accelerated to the nearest electrode which they pass through as the alternating p.d. reverses polarity. The particles are then repelled on leaving this electrode and attracted to the next so the procedure is repeated and the charged particles gain kinetic energy each time they pass between the electrodes. No magnetic field is needed as the charged particles do not change direction. The first direct evidence for quarks was discovered from the Stanford Linear Accelerator.

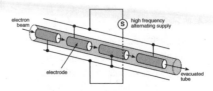

The linear accelerator

see also...

Particle Interactions; Quarks 1 and 2

Particle Interactions

The four fundamental forces are gravity, electromagnetism, the strong nuclear force and the weak nuclear force. These forces act by means of the exchange of packets of energy referred to as quanta. Diagrams used to represent these interactions were first devised by Richard Feynman and are known as Feynman diagrams.

★ The electromagnetic force includes static electricity and magnetism and acts between charged particles as a result of the exchange of massless quanta known as photons. These photons are referred to as 'virtual' because the interaction would cease if a detector was used to detect them.

★ The strong nuclear force holds neutrons and protons together in the nucleus. Protons and neutrons consist of three fundamental particles called quarks, held together as a result of the exchange of quanta known as gluons. A gluon in a proton or neutron can produce a quark–antiquark pair such that the antiquark and a different quark form a composite particle known as a pion which crosses to a different proton or neutron. This 'pion exchange' process is the mechanism through which the strong nuclear force acts.

★ The weak nuclear force causes a proton to turn into a neutron in a proton-rich nucleus or a neutron into a proton in a neutron-rich nucleus. In this process, a short-lived particle called a W boson is created. In β⁻ decay, a neutron changes into a proton and emits a W⁻ boson that decays into a β⁻ particle (i.e. an electron) and an antineutrino. In β⁺ decay, a proton changes into a neutron and emits a W⁺ boson which decays into a positron and a neutrino.

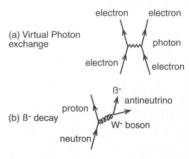

(a) Virtual Photon exchange

electron   electron

photon

electron   electron

(b) β⁻ decay

β⁻

antineutrino

proton

W⁻ boson

neutron

Feynman diagrams

see also...

Photon; Quarks; Radioactivity 2

Photoelectricity

The photon theory of electromagnetic radiation was established by Einstein to explain the photoelectric effect which is the emission of electrons from a cold metal when light above a certain frequency is directed at the metal surface. This effect was discovered in 1888 by Hallwachs who found that an insulated negatively charged zinc plate discharged itself when it was illuminated with ultraviolet light. Further investigations showed that the metal emits electrons when illuminated and that the effect did not occur if the light frequency was less than a threshold value, no matter how intense the light was.

Every metal contains conduction electrons which are electrons free to move about inside the metal. When a metal is heated, these electrons gain kinetic energy and move about faster inside the metal. Conduction electrons at the surface of a metal can also gain kinetic energy from light directed at the surface. The existence of the threshold frequency could not be explained by the wave theory of light which predicted that light of any frequency should be capable of causing photoelectric emission.

In 1905, Einstein put forward the new idea that light consists of wavepackets of electromagnetic energy which he called photons. He assumed that the energy E of a photon was proportional to its frequency, f, in accordance with the equation $E = hf$ where h is the Planck constant. When light is directed at a metal, electrons in the metal near the surface absorb photons. Each electron that absorbs a photon thus gains kinetic energy equal to the photon's energy (hf). To leave an uncharged metal surface, each electron needs a minimum amount of energy which is referred to as the work function of the metal ϕ. The kinetic energy of a conduction electron is negligible until it absorbs a photon. Thus an electron that absorbs a photon can leave the metal surface if the energy of a single photon exceeds the work function (i.e. $hf > \phi$). Thus photoelectric emission can only occur from an uncharged metal plate if the light frequency exceeds ϕ/h, which is referred to as the threshold frequency of the metal.

see also...

Electromagnetic Waves; Photon

Photon

A photon is a packet of electromagnetic waves of energy $E = hf$, where f is the frequency of the electromagnetic waves. A photon is the quantum or least quantity of electromagnetic radiation from a source of electromagnetic radiation of well-defined frequency.

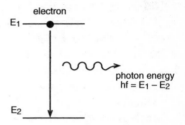

An electron transition

When an electron moves from an outer shell to a vacancy in an inner shell of an atom, it loses energy which it releases as a photon of electromagnetic energy. If an electron transfers from an energy level E_1 to a lower energy level E_2, the energy of the photon released = $hf = E_1 - E_2$. For light atoms, the photons are in the optical or ultraviolet range. X-ray photons are emitted when deep inner-shell vacancies in large atoms are filled. Gamma photons are emitted when nuclei with excess energy de-excite to the ground state.

For a point source of photons emitting energy at a rate W, the number N of photons per second emitted by the source = W/hf since each photon carries energy hf. Hence the number of photons per second passing at right angles through unit area at a distance r from the source = $(W/hf)/4\pi r^2$. The intensity at distance r from the source = energy per second passing through a unit area at right angles to the area = $W/4\pi r^2 = Nhf/4\pi r^2$. The intensity therefore varies with distance according to the inverse square law. This inverse square law applies to the light emitted by a star and to the intensity of radiation from a source of γ radiation. In both cases, no radiation is absorbed so the intensity decreases with increased distance because the radiation spreads out in all directions.

see also...

Energy Levels in Atoms; Inverse Square Laws

Polarisation

Transverse waves are waves whose vibrations are perpendicular to the direction of propagation of the waves. Examples include electromagnetic waves, waves on a vibrating string and secondary seismic waves. Transverse waves are plane polarised if they vibrate in one plane only. Unpolarised transverse waves vibrate in a randomly changing plane. Longitudinal waves are waves whose vibrations are parallel to the direction of propagation of the waves. Examples include sound waves and primary seismic waves.

Electromagnetic waves are transverse waves as they consist of oscillating electric and magnetic fields at right angles to each other and to the direction of propagation. The electric field of a polarised electromagnetic wave vibrates in one plane only and the magnetic field vibrates in a perpendicular plane. The plane of polarisation of a polarised electromagnetic wave is defined as the plane of vibration of its electric field.

Sunlight and light from a filament lamp or a flame is unpolarised and can be polarised by passing it through a polaroid filter. The filter molecules allow light through only if the plane of vibration of the light waves is at right angles to the alignment of the filter molecules. If polarised light is incident on a second polaroid filter, the intensity of the light transmitted by the second filter is greatest if the filters are aligned so their molecules are perpendicular to the plane of vibration. If the two filters are perpendicular to each other, no light passes through the second filter.

Unpolarised light is polarised when it reflects off glass or water. Polarisation by reflection is complete at a certain angle of incidence and partial at all other angles. This is why polaroid sunglasses eliminate glare when sunlight reflects off water. The polaroid sunglasses prevent light reflected off the water surface from passing through but allow through light from elsewhere.

see also...

Electromagnetic Waves; Liquid Crystals

Potential Difference and Power

★ Potential difference (p.d.) is the potential energy gained or lost per unit charge by a small positive charge when it moves from one point to another point. The word 'voltage' is commonly used instead of p.d. The potential energy of a charge is usually referred to as electrical energy.

Potential difference $V = \dfrac{E}{Q}$

where E = energy delivered, and Q = charge passed.

★ The unit of potential difference is the volt (V) which is the p.d. between two points if one joule of electrical energy is gained or lost when one coulomb of charge passes between the two points.

★ The e.m.f. of a source of electrical energy is the gain of electrical energy per unit charge of a small charge that passes through the source. In a circuit, the flow of charge round the circuit transfers energy from the sources of e.m.f. to the components in the circuit.

★ The potential drop across a component is the loss of electrical energy per unit charge of a small charge that passes through the component. The potential drop across a component in an electric circuit is like the pressure drop between the inlet pipe and the outlet pipe of a radiator in a central heating system. The pressure difference is necessary to drive water through the radiator.

Electrical power is defined as the electrical energy transferred per second to or from part of an electric circuit. The unit of power is the watt (symbol W). One watt is defined as a rate of transfer of energy of one joule per second: 1 kilowatt = 1000 watts.

Because current is charge per second flowing through a component or device, and potential difference is the electrical energy delivered per unit charge to the component or device, therefore

current × potential difference =

$\dfrac{\text{charge}}{\text{time}} \times \dfrac{\text{electrical energy}}{\text{charge}} =$

$\dfrac{\text{electrical energy}}{\text{time}}$ = power.

Mains electricity is priced in 'units' of kilowatt hours (kW h) where 1 kW h is the electrical energy delivered in 1 hour at a rate of 1 kilowatt.

see also...

Charge and Current

69

Pressure

The pressure on a surface is defined as the force per unit area acting at right angles to a surface. The unit of pressure is the pascal (Pa) which is equal to 1 newton per square metre.

Pressure =
<u>force acting normally to a surface</u>
surface area

★ In a liquid at rest, the pressure at a certain depth acts equally in all directions and is the same at all points at the same depth in the liquid. The pressure increases with depth in accordance with the equation $p = h\rho g$ where h is the depth below the surface and ρ is the density of the liquid.

★ In a gas at rest, the pressure on the container surface is due to countless impacts of fast-moving molecules hitting and rebounding from the surface. The higher the temperature of a gas in a sealed container, the greater the pressure because the gas molecules move faster and hit the surface harder and more often. The ideal gas laws can be explained using the kinetic theory of gases.

★ In a liquid or a gas in motion, the pressure on a surface at right angles to the flow exceeds the pressure on a surface parallel to the flow. The pressure on a surface parallel to the flow is called the static pressure as it is the same as the pressure at that point if the fluid was at rest. The pressure on a surface perpendicular to the flow is referred to as the total pressure. The difference between the pressure on a surface perpendicular to the flow and on a surface parallel to the flow is called the dynamic pressure.

Atmospheric pressure varies slightly from one day to the next according to the state of the weather. On average at sea level, atmospheric pressure is 101 kPa. This is referred to as standard pressure. Atmospheric pressure lessens with height thus causing breathing difficulties for mountaineers at very high altitudes. Pressure gauges are usually calibrated to measure the difference between the pressure of a gas or liquid and atmospheric pressure.

> ## see also...
> *Fluids 1 and 2; Ideal Gases*

Projectiles

A projectile is any object in motion acted on only by the Earth's gravity. At any point on its flight path, its horizontal component of acceleration is zero and its vertical component of acceleration is equal to g, the gravitational field strength at that point.

A projectile's vertical motion is unaffected by its horizontal motion. The trajectory of a projectile can be calculated using the dynamics equations for constant acceleration.

(i) If a projectile is released from rest, it has no horizontal motion. Its speed increases at a constant rate equal to g, the acceleration of an object in free fall. Thus at time t after being released:

★ its speed $v = gt$,

★ its average speed during the period from release to time t after release $= gt / 2$,

★ its loss of height, h = average speed × time taken $= gt^2 / 2$.

(ii) If a projectile is launched horizontally at speed U, then at time t after being launched:

★ its horizontal distance from the launch point $x = Ut$ since it moves at constant speed horizontally,

★ its vertical motion is the same as that of an object released from rest at the same instant, hence its loss of height $h = gt^2 / 2$.

The flight path of an object launched horizontally is therefore a curve which becomes steeper and steeper as it descends. This type of curve is referred to as a parabolic curve.

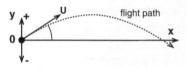

A projectile

(iii) If the projectile is projected in an initial direction above the horizontal, it moves equal distances horizontally in equal times. Its vertical speed decreases to zero and it then falls at increasing speed. Its flight path is a a parabolic curve which is symmetrical about the highest point.

see also...

Dynamics

Quantum Theory

Quantum theory is about physical quantities such as charge and energy which exist only in multiples of a basic amount, referred to as a quantum. The quantum theory was devised by Planck in 1900 to explain the spectrum of thermal radiation from a hot object. The intensity of radiation from a hot object varies continuously with wavelength, reaching peak intensity at a certain wavelength. The classical theory of radiation established in 1848 could not explain the peak of the radiation curve as it predicted the intensity becomes infinite as the wavelength became smaller and smaller. This prediction was known as the Ultraviolet Catastrophe. Planck was able to explain the shape of the radiation curve by assuming that the energy of each atom of the radiation source is quantised in multiples of a basic amount, hf, where f is the frequency of vibration of the atom. In addition, he assumed that the energy of an atom could change only by one quantum of energy (= hf) when the atom absorbed or emitted radiation. Planck's quantum theory avoids the Ultraviolet Catastrophe because the shorter the wavelength of the radiation, the higher its frequency and so the higher the energy levels of the atoms vibrating at this frequency. Fewer and fewer such atoms would be above the lowest energy level. Thus the intensity of radiation emitted would fall to zero for shorter and shorter wavelengths.

The theory that electromagnetic radiation is quantised was put forward by Einstein to explain the photoelectric effect. He assumed that the quantum of electromagnetic radiation which he named the photon has energy equal to hf, where h is the Planck constant and f is the frequency of the radiation.

Electric charge is also quantised. In 1915, Robert Millikan showed that the charge of an oil droplet is always an integer multiple of a basic unit of charge which he assumed to be the charge of the electron.

see also...

Electron; Photon; Wave Particle Duality

Quarks 1

Quarks are the building blocks of protons and neutrons. There are six different quarks. Three of them carry a charge of $+^2/_3$ e. The other three carry a charge of $-^1/_3$ e:

1 The up quark ($+^2/_3$ e) and the down quark ($-^1/_3$ e) make up protons and neutrons.

2 The charmed quark ($+^2/_3$ e) and the strange quark ($-^1/_3$ e) are heavier than the up and down quarks and are unstable.

3 The top quark ($+^2/_3$ e) and the bottom quark ($-^1/_3$ e) are the heaviest quarks and the least stable.

The quark model explains all known baryons, antibaryons and mesons. Mesons and baryons are known collectively as hadrons.

★ A baryon consists of three quarks. An antibaryon consists of three antiquarks. For example, a proton consists of two up quarks and a down quark (i.e. uud) and a neutron consists of one up quark and two down quarks (i.e. udd).

★ A meson consists of a quark and an antiquark. For example, a pion or π meson consists of an up or a down quark and an up or down antiquark.

The first evidence for quarks was obtained when it was discovered that high energy electrons in a beam were scattered from a stationary target by three scattering centres in each proton or neutron. The beam of electrons was produced in the Stanford Linear Accelerator at sufficiently high energy to probe the sub-structure of protons and neutrons. The results supported the quark model which had been put forward by Murray Gell Mann and Georg Zweig as an explanation of the patterns of particles discovered in high energy collisions between hadrons.

Quarks do not exist in isolation. In a high energy collision, pair production of quarks and antiquarks takes place when hadrons collide. As a result, more hadrons are produced and no quark or antiquark remains outside a hadron. The quarks in a hadron move about relatively freely provided they do not move away from each other. The force between the quarks in a hadron is due to the exchange of gluons.

see also...

Particle Accelerators; Particle Interactions

Quarks 2

Matter consists of fundamental particles which are either leptons (e.g. electrons, positrons, neutrinos and antineutrinos), or quarks.

Matter particles were originally classified in three groups according to mass:
1 particles lighter or as light as the electron, referred to as leptons;
2 particles heavier or as heavy as the proton, referred to as baryons;
3 particles lighter than protons and heavier than electrons, referred to as mesons.

The particles in each group differ in terms of exact mass, charge, lifetime and strangeness. This last property was discovered when it was observed that certain particles were produced in pairs as a result of a strong interaction, and were found to decay through the weak interaction. Strangeness was introduced as a quantum number which is conserved in the strong interaction.

As a result of classifying baryons and mesons in each group according to charge, strangeness and lifetime, it was deduced that each baryon consists of three quarks, each

antibaryons consists of the three corresponding antiquarks and each meson is a quark and an antiquark.

Figure 13 shows the possible combinations of quarks and antiquarks that make up baryons and mesons respectively.

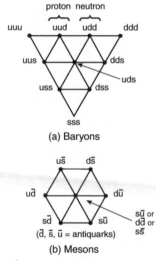

Quark patterns

Leptons are thought to be elementary particles, not composite particles.

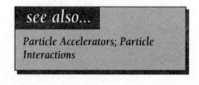

see also...

Particle Accelerators; Particle Interactions

Radioactivity 1 – Radioactive Decay

A radioactive isotope decays as a result of one of the following nuclear transformations:

1 Alpha emission occurs when a large unstable nucleus emits two protons and two neutrons as a single particle, referred to as an alpha (α) particle.

$$^A_Z X \rightarrow {}^4_2 \alpha + {}^{A-4}_{Z-2} Y$$

2 Beta minus emission occurs when a neutron in a neutron-rich nucleus changes into a proton.

$$^A_Z X \rightarrow {}^0_{-1}\beta + {}^A_{Z+1} Y$$

3 Gamma emission, in which a gamma photon is emitted from a nucleus which has excess energy after emitting an alpha or a beta particle.

The theory of radioactive decay is based on the assumption that it is a random process and that the probability of a nucleus disintegrating in a given time interval Δt is proportional to Δt. Hence:

$$\frac{dN}{dt} = -\lambda N,$$

where λ is the decay constant. The solution of this equation is $N = N_0 e^{-\lambda t}$, where N_0 is the initial number of atoms present.

The activity of a radioactive isotope is the number of nuclei of the isotope that disintegrate per second. Hence activity $A = dN/dt$, where N is the number of radioactive nuclei remaining. Since $A = -\lambda N$, then the activity A of a radioactive isotope also decreases exponentially, in accordance with the equation $A = A_0 e^{-\lambda t}$, where A_0 is the initial activity.

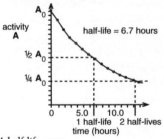

A half-life curve

The **half-life** of the isotope is the time taken for the number of nuclei of the isotope to decrease to 50%. This is the same as the time taken for the activity to decrease to 50%. Since $N = 0.5N_0$ after one half-life, then $0.5N_0 = N_0 e^{-\lambda t}$ which gives $\lambda T_{1/2} = \ln 2$.

see also...

Nuclear Model of the Atom; Radioactivity 2

Radioactivity 2 – Properties of α, β and γ Radiation

1 Range in air at atmospheric pressure:

α *radiation*; the alpha particles from a particular isotope are emitted with the same kinetic energy which varies from one isotope to another. The range in air of the alpha particles from a particular isotope is therefore well-defined and can be up to 10 cm.

β *radiation*; the beta particles from a particular isotope possess a continuous spread of initial kinetic energies up to a maximum that depends on the particular isotope. The range in air varies up to about a metre.

γ *radiation*; gamma photons spread out from a point source in all directions, interacting little with air molecules. The range in air is unlimited although the intensity of gamma radiation from a point source varies with distance from the source according to the inverse square law because the photons spread out in all directions without absorption.

2 Absorption by matter:

α *radiation*; alpha particles are absorbed by paper or thin card or thin metal foil.

β *radiation*; beta particles penetrate paper, thin card and thin metal foil. Aluminium plates of thickness more than about 5 mm absorb beta particles.

γ *radiation*; gamma photons are absorbed by lead plates of thickness about 50 mm.

3 Ionisation:

(i) α particles produce of the order of 10 000 ions per millimetre in air which is much more than β particles or γ radiation.

(ii) β particles produce fewer ions per millimetre than α particles do because they are much faster than α particles.

(iii) γ radiation produces very little ionisation in air as photons are uncharged. Note that X-rays are also high energy photons and therefore produce ionisation. X-rays are produced in an X-ray tube.

---

see also...

Ionisation; Inverse Square Laws; Radioactivity 1

Radioactivity 3 – Radioactive Waste

Radioactive waste is classed as low level, medium level and high level.

★ Low level waste consisting of equipment and clothing used by radiation workers is buried in sealed containers at supervised sites. Cooling water used in the heat exchangers in nuclear power stations is slightly radioactive and is discharged into the sea.

★ Medium level waste such as the coolant used in a nuclear power station is solidified and stored in sealed containers in underground chambers at supervised sites.

★ High level waste consists of control rods, fuel containers and spent fuel after it has been reprocessed. The waste is stored for many years in sealed containers in underground chambers at supervised sites.

The spent fuel rods from a nuclear reactor contain unused uranium 235, uranium 238, plutonium 239 produced as a result of uranium 238 absorbing neutrons, and neutron-rich fission fragments. The various isotopes in a spent fuel rod therefore include α, β and γ emitters with a wide range of half-lives. The shortest half-life isotopes have the greatest activity per unit mass which is why a spent fuel rod on removal from a reactor is highly radioactive. The spent fuel rods are handled by remote control from removal to reprocessing to storage.

On removal from a reactor, a spent fuel rod is placed in a cooling pond for up to a year until its short-life isotopes have decayed and its rate of release of energy is no longer a problem. The fuel container is then cut open and the spent fuel is removed and reprocessed chemically and the unused uranium and plutonium is recovered and stored for future use. All the other material is stored in sealed containers as high level waste. Long term corrosion of the containers is a potential problem which may be solved by vitrification. This process involves mixing the material with molten glass and allowing it to solidify in glass blocks which do not corrode the containers. Care must be taken not to store excess quantities of uranium and plutonium in close proximity, otherwise fission might start and an explosion could result.

see also...

Fission; Nuclear Power

Radioactivity 4 – Measurement of Ionising Radiation

Ionising radiation kills living cells as a result of damaging cell membranes beyond repair and also destroys the mechanism of replication in cells as a result of damaging the DNA strands in cell nuclei. In addition, ionising radiation creates free radicals which can cause tumours.

★ The radiation dose received by a substance from ionising radiation is the energy deposited per unit mass in the substance. The unit of radiation dose is the gray (Gy) which is equal to 1 J kg^{-1}.

★ The relative biological effectiveness (r.b.e.) of a given form of ionising radiation is the radiation dose needed to achieve a certain biological effect with 250 kV X-rays in comparison with the given form of radiation. For example, α-radiation has a r.b.e. of 10 which means that ten times the dose of 250 kV X-rays is needed to produce the same effect as a given dose of α-radiation.

★ The dose equivalent received by a living substance from a certain dose of ionising radiation is the dose of 250 kV X-rays needed to give the same biological effect as the ionising radiation. The unit of dose equivalent is the sievert (Sv) which is equal to 1 J kg^{-1}.
Dose equivalent = radiation dose × relative biological effectiveness.

★ The total dose equivalent for exposure to different types of radiation is the sum of the dose equivalents for each type of radiation.

No lower limit exists for biological damage by ionising radiation and maximum permissible exposure limits to ionising radiation are determined by law in accordance with what is considered to be acceptable in terms of risk. The maximum permissible limits in the UK is 15 mSv per annum for occupations where ionising radiation in used and 0.5 mSv per annum above naturally occuring ionising radiation levels for the general public. These limits are based on estimates of three fatal cancers per millisievert per 100 000 survivors of the atomic bombs dropped on Hiroshima and Nagasaki. Thus a limit of 0.5 mSv per annum corresponds to 750 deaths per year in a population of 50 million people.

see also...

Ionisation; Radioactivity 1, 2 and 3; X-rays

Random Processes

A random process is a change or an event which is unpredictable. The statistical outcome of a very large number of random changes or events is very predictable. The two examples below are chosen to illustrate this point.

★ Radioactive decay of an unstable isotope; the decay of an unstable nucleus is a random event. Therefore each unstable nucleus is as likely to disintegrate in a given time interval as any other unstable nucleus of the isotope. Thus the probability of the decay of a nucleus in a certain time interval is the same for any nucleus.

Hence, for N unstable nuclei, the number of nuclei ΔN that disintegrate in a time interval Δt is proportional to N and to Δt, i.e. ΔN is proportional to $N \Delta t$. Thus the percentage of nuclei (= $\Delta N / N$ x 100%) that decay in a given time is proportional to the time interval Δt. Thus the number of nuclei decreases exponentially.

★ Diffusion; the molecules in a gas or a liquid are in continual random motion, repeatedly colliding with each other and with the container walls. Air molecules move with a continuous range of speeds of the order of hundreds of metres per second. If an air freshener is sprayed into a corner of a room, the scent takes in the order of minutes to reach all parts of the room because the scent molecules gradually spread from high concentration to low concentration, their progress repeatedly impeded by impacts with each other and with air molecules. The progress of a scent molecule from its starting point is not unlike the progress of a person on a gigantic chessboard who steps from square to adjoining square at random. After 100 steps, such a person would on average be 10 steps from the start, after 400 steps just 20 steps from the start. The theory of such a two-dimensional random walk leads to the result that if it is repeated many times, the person is most likely to be $N^{1/2}$ steps from the start after N steps. The same idea can be applied to the random progress of a molecule in a gas or a liquid.

see also...

Decay Processes; Radioactivity 1

Red Shift

The **Doppler effect** is the change of the observed frequency of waves from a source due to relative motion between the source and the observer. The change of frequency is referred to as the **Doppler shift**. The Doppler effect is used in a wide range of applications including radar speed traps and blood flow measurements.

Consider a small moving source of waves emitting waves at a constant frequency. The waves spread out from the moving source, bunching together ahead of the source and spacing out behind the source. An observer ahead of the source would detect a reduced wavelength and an observer behind the source would detect an increased wavelength.

★ For sound waves, the observed frequency = speed of the sound waves / wavelength.
(i) An observer ahead of the source would detect a higher frequency provided the distance between the source and observer is decreasing.
(ii) An observer behind the source would detect a lower frequency,
provided the distance between the source and observer is increasing.
★ For electromagnetic waves, the observed frequency = speed of light / wavelength. The speed of light is the same for any observer. If the source is moving away from an observer, the wavelength is increased towards the red part of the spectrum, referred to as a red shift. If the source is moving towards the observer, the wavelength is reduced towards the blue part of the spectrum, a blue shift. If the change of wavelength is measured, the speed of the light source can be calculated from the equation speed of source / speed of light = change of wavelength / wavelength from a stationary source.

see also...

Electromagnetic Waves; Hubble's Law; Wavemotion 1 and 2

Resistance

The resistance of a component or a wire is a measure of the difficulty of making the component conduct electricity. Its resistance is defined as potential difference/current. The unit of resistance is the ohm (symbol Ω). One ohm is the resistance of a conductor through which the current is one ampere when the potential difference across the conductor is one volt.

Ohm's Law states that the resistance of a wire at constant temperature is independent of the current through it. A graph of potential difference (on the y-axis) against current for an ohmic conductor is a straight line because its resistance is constant. A filament light bulb's resistance increases as the current increases so the filament is non-ohmic.

For resistors R_1, R_2, R_3, etc.
(i) in series, their combined resistance $R = R_1 + R_2 + R_3 + \cdots$.
(ii) in parallel, their combined resistance R is given by $1/R = 1/R_1 + 1/R_2 + 1/R_3 + \cdots$.

Internal resistance: The electrical energy produced by a source of electricity in a circuit is delivered to the components of the circuit by charge flowing round the circuit. Some of this energy is wasted due to the source's internal resistance. The electromotive force (e.m.f.) of an electrical source is the electrical energy produced per unit charge by the source. The lost p.d. in the source due to its internal resistance is the electrical energy wasted per unit charge inside the source. The lost p.d. depends on the current and on the internal resistance of the source.

For a source of e.m.f. E and internal resistance r connected to a load of resistance R, the load p.d. falls as the current I increases because the load p.d. $IR = E - Ir$. This is why the output potential difference of a source of electrical energy (including a power supply unit) falls if more current is drawn from the source. In an old house, the lights can dim when you switch on an electric cooker.

see also...

Potential Difference and Power

Resonance

Resonance occurs when a periodic force is applied to an oscillating system and the amplitude of the oscillating system is very large as a result. In any system which oscillates freely, kinetic energy is changed to potential energy and back every half-cycle. If frictional forces are present and no periodic force is applied, the amplitude of oscillations becomes less and less and the total energy of the system decreases as energy is transferred to the surroundings by the frictional forces. The oscillations are said to be 'damped' by the presence of frictional forces. For a lightly damped system which is subjected to a periodic force, resonance occurs when the frequency of the periodic force is equal to the frequency of vibration of the system. The applied force is in phase with the system at resonance. The energy supplied by the periodic force is equal to the energy lost due to the frictional forces at resonance.

A simple example is provided by a child on a swing who is pushed periodically. If the frequency of the pushes is equal to the natural frequency of oscillation, f_0, the amplitude becomes very large, limited only by the frictional forces present. The frequency at which the amplitude is greatest is called the resonant frequency and is equal to f_0 for light damping.

Examples of resonant systems include:
★ mechanical resonance, when a panel of a washing machine vibrates loudly at a certain motor speed;
★ acoustic resonance, when air is blown at a certain speed over the top of an empty bottle making the bottle resonate with sound;
★ electrical resonance, when a radio is tuned to a certain station as a result of adjusting the frequency dial to the station frequency so that radio waves of that frequency cause a sufficiently large alternating p.d. at that frequency across the tuning circuit.

see also...

Simple Harmonic Motion;
Wavemotion 2

Satellite Motion

Any object in orbit about a larger object is a satellite. The planets are satellites of the Sun. The Moon is a satellite of the Earth. Artificial satellites in orbit about the Earth are used for communications.

★ The time period of a satellite is the time taken to complete one orbit. The time period of a satellite depends on its height.

★ A geostationary satellite is a satellite in an equatorial orbit at such a height and direction that it remains at the same point above the equator because its time period is exactly 24 hours.

A satellite orbiting the Earth is kept on its orbit by the force of gravity between it and the Earth. In general, a satellite orbit is an ellipse. For a circular orbit, the velocity of the satellite is always perpendicular to the force of gravity on the satellite, and is given by equating the force of gravitational attraction GMm / r^2 to the centripetal force $m\upsilon^2 / r$, where M is the mass of the Earth, m is the mass of the satellite, r is the radius of orbit and υ is the speed of the satellite. Hence $\upsilon^2 = GM / r$ gives the speed of a satellite in a circular orbit.

The time period $T = 2\pi r / \upsilon$ for a circular orbit, hence

$$T^2 = \frac{4\pi^2 r^2}{\upsilon^2} = \frac{4\pi^2 r^2}{(GM/r)} = \frac{4\pi^2 r^3}{GM}$$

This equation agrees with Kepler's 3rd Law which states that T^2 is proportional to r^3 for the planets. The equation above can be used to show that $T = 24$ hours requires $r = 42\ 300$ km, corresponding to a height of 35 900 km above the Earth. Thus the height of a geostationary orbit must be 35 900 km because the Earth's radius is 6400 km. A geostationary satellite is said to be in synchronous orbit. It remains at the same point above the equator because it moves round its orbit at the same rate as the Earth spins. Geostationary satellites are used for communications because satellite transmitters and receivers need no further positional adjustment once pointed towards the satellite.

> ### see also...
> *Circular Motion; Gravitational Fields 1*

Simple Harmonic Motion

An oscillating object moves to and fro along a line.

★ the amplitude of its motion is its maximum displacement from the centre of the oscillations;

★ the time period, T_p, of its motion is the time it takes to complete one cycle of oscillation which is from one extreme to the other and back again.

The motion of an oscillating object is simple harmonic motion (s.h.m.) if its acceleration is in proportion to its displacement from the centre of the oscillations. This condition may be expressed as an equation of the form 'acceleration = − constant × displacement' where the minus sign is because the acceleration always acts towards the centre whereas the displacement is always measured from the centre. The constant of proportionality in this equation is the square of the angular frequency ω which is $2\pi/T_p$. Thus the essential condition for s.h.m. is that the acceleration a and the displacement s must fit the equation $a = -\omega^2 s$. Clearly, the acceleration of the object is at its greatest magnitude when the object is furthest from the centre of oscillations.

In a system where an object of mass m oscillates because of the presence of one or more springs, the force restoring the object to equilibrium depends on the stretching of the springs. A spring system obeys **Hooke's Law**, namely the tension in a spring = ke, where e is the extension of the spring and k is the spring constant. Thus the restoring force $F = -ks$ for displacement s from equilibrium. Using Newton's 2nd Law in the form $F = ma$ therefore gives $a = F/m = -(k/m)s$. Hence the motion is s.h.m. and $k/m = \omega^2$. Therefore the time period $T_p = 2\pi/\omega = 2\pi(m/k)^{1/2}$.

If either the mass is increased or the springs become weaker, the time period is therefore increased so the oscillations take longer. Any mechanical system where one or more springs cause oscillations has a time period given by the above equation.

see also...

Force and Motion

Special Relativity 1 – Principles

Relativity has to do with relative motion and the failure to detect absolute motion. In 1905, Albert Einstein put forward the theory of relativity, now known as special relativity, to explain why absolute motion cannot be detected. Einstein's Special Theory of Relativity is based on two postulates:

1 The speed of light in free space, c, is invariant which means it is the same, regardless of the speed of the light source or of any observer.

2 All physical laws in the form of equations can be expressed in the same form in any inertial frame of reference. An inertial frame of reference is a frame of reference in which an object at rest remains at rest, provided no forces act on it.

Einstein assumed the invariance of c as a starting point. He considered two coordinate systems, one (O') moving at speed v along the x-axis relative to the other (O). At the instant the origins coincide, a light wave is emitted from this point.

1 The distance r moved by the light wave in time t in coordinate system O is given by $r = ct$, where $r^2 = x^2 + y^2 + z^2$, therefore $x^2 + y^2 + z^2 = c^2 t^2$.

2 The distance r' moved by the light wave in time t' in coordinate system O' is given by $r' = ct'$, where $r'^2 = x'^2 + y'^2 + z'^2$, therefore $x'^2 + y'^2 + z'^2 = c'^2 t'^2$.

Since the motion of O' relative to O is along the x-axis, then y and z are unaffected so $y = y'$ and $z = z'$, therefore $y^2 + z^2 = c^2 t^2 - x^2 = c'^2 t'^2 - x'^2$. With the condition $c^2 t^2 - x^2 = c'^2 t'^2 - x'^2$ to be met, Einstein worked out that
$x' = \gamma(x - vt)$ and
$t' = \gamma(t - vx/c^2)$
where the Lorentz factor
$\gamma = (1 - v^2/c^2)^{-1/2}$.

These equations are known as the Lorentz transformation equations. The consequences of these equations include time dilation, length contraction, relativistic mass and $E = mc^2$.

see also...

Special Relativity 2 and 3

Special Relativity 2 – Length Contraction and Time Dilation

The Lorentz equations can be used to show that:

★ the observed length l of a moving rod $= l_0 / \gamma$, where υ is the speed of the rod, γ is the Lorentz factor $(1 - v^2/c^2)^{-1/2}$ and l_0 is the proper length of the rod which is its length as measured by an observer who is stationary with respect to the rod. Because γ is greater than 1 for any moving object, then the observed length is always shorter than the proper length,

★ the time interval t between the two events as measured by an observer moving at constant velocity υ relative to the events is stretched out or 'dilated' in accordance with the equation $t = \gamma \, t_0$, where t_0 is the proper time interval as measured by an observer who is stationary relative to the events. Because γ is greater than 1 for any moving object, then the time interval is always more than the proper time.

Experimental evidence for time dilation and length contraction was found in experiments involving high energy unstable particles which move at speeds approaching the speed of light. Such experiments have been carried out on unstable particles called muons. Measurements on muon intensities in the upper atmosphere and at ground level show that most of the muons created in the upper atmosphere reach ground level 2 km below. However, the 'proper' half-life of a muon is known to be about 1.5 μs which means that most of the muons should decay after travelling 2 km. The large discrepancy is explained by time dilation. The half-life of the muons produced by cosmic radiation is stretched because these muons are travelling at almost the speed of light so they last much longer than stationary muons do.

An observer moving at the same speed as these cosmic muons would see them decay at their normal short-lived rate but the atmosphere would appear contracted so the fraction reaching the ground would be the same.

see also...

Special Relativity 1 and 3

Special Relativity 3 – Mass and Energy

instein proved in his 1905 papers on relativity that the mass of an object depends on its speed and that if an object is supplied with energy, its mass increases and if energy is removed from it, its mass decreases.

★ Mass is characterised by inertia which is the difficulty of changing the motion of an object. Einstein proved that the mass m of an object depends on its speed v in accordance with the equation $m = \gamma m_0$ where m_0 is the rest mass of the object and γ is the Lorentz factor $(1 - v^2/c^2)^{-1/2}$.

★ Energy is the capacity to do work. If energy ΔE is transferred to or from an object, Einstein showed that its mass changes by an amount Δm in accordance with the equation $\Delta E = \Delta mc^2$, where c is the speed of light in free space. For any object of mass m, its total energy $E = mc^2$.

The change of mass due to energy change is not significant in chemical reactions and in gravitational changes near the Earth.

(i) A 1 kg mass would need to gain 64 MJ of potential energy to leave the Earth completely. This would increase the mass of the 1 kg object and of the Earth by an insignificant amount.

(ii) A typical chemical reaction involves energy change of the order of an electron volt (= 1.6×10^{-19} J). The mass change due to such an energy change is much much smaller than the mass of an electron.

(iii) Change of mass due to energy change is significant in nuclear reactions where extremely strong forces confine protons and neutrons to the nucleus, overcoming the electrostatic repulsion of the protons except when an unstable nucleus disintegrates. Nuclear energy changes are typically of the order of MeV per nucleon, about a million times larger than chemical energy changes. The change of mass for an energy change of 1 MeV is therefore not insignificant compared to the rest mass of a nucleon. The mechanism which causes the mass of an object to change when its energy changes is not yet clear, even though there is abundant experimental evidence for the equation $E = mc^2$.

see also...

Special Relativity 1 and 2

States of Matter

The physical state of substances is usually solid, liquid or gaseous. Some substances can exist in a physical state that cannot be neatly classed in this way.

(i) A solid has a definite shape and it has a surface. The atoms and molecules of a solid are held together in fixed positions by bonds between them. The atoms vibrate about their mean positions. Raising the temperature of a solid increases the energy of vibration of the atoms. If a pure solid is heated, it melts at a certain temperature. The energy supplied at the melting point is used to enable the atoms of the solid to break free from each other, overcoming the strong bonds that lock them together in the solid state.

(ii) A liquid has a surface, can flow and it takes the shape of its container. The molecules in a liquid move about at random in contact with each other. Intermolecular forces prevent the molecules from moving away from each other so retaining the open surface of the liquid. Raising the temperature of a liquid increases the kinetic energy of the molecules, enabling them to move about faster. If a pure liquid is heated, it boils at a certain temperature. The energy supplied enables the molecules to move away from each other and enter the space above the liquid surface. Before reaching boiling point, molecules at the surface with sufficient kinetic energy can break free from the other liquid molecules to enter the space above the liquid. This process of evaporation increases as the temperature increases.

(iii) A gas can flow but it has has neither shape nor surface. The critical temperature of a gas is the temperature above which it cannot be liquefied by compression. Below this temperature, the term 'vapour' is used for the gaseous state as it can be liquefied below its critical temperature if it is compressed sufficiently at a constant temperature or if it is cooled sufficiently at constant pressure. A saturated vapour is where the vapour is in thermal equilibrium with the liquid so the space above it cannot contain any more vapour molecules.

see also...

Ideal Gases; Structure of Materials

Structure of Materials

Solids are either crystalline (atoms are arranged in a regular pattern), polymeric (linked together in long chain molecules) or amorphous (arranged at random). Composite materials consist of two or more materials combined together.

★ Crystalline solids include metals as well as crystals. Because the atoms in a crystal are arranged in a regular pattern on such a large scale, the surfaces of a crystal form well-defined angles to each other. A metal consists of many tiny crystals called grains packed together with no spaces between. The atoms in each grain form a regular pattern. The grains are aligned at random to each other.

★ Ceramic materials consists of lots of tiny crystals or grains locked together in a glassy cement such as silica. Ceramics are chemically stable because the outer electrons in the atoms of the glassy substance are held firmly in the strong bonds between the atoms of the substance so these electrons cannot interact with ions from other substances. The very high melting points of ceramics are because the crystals in them are

ionic and so the ions are held together by strong ionic bonds.

★ Amorphous solids consist of atoms or groups of atoms joined together at random in a rigid structure. Glass is an amorphous solid. No regularity exists in the arrangement of the atoms in an amorphous solid. 'Amorphous' means 'without shape'.

★ Polymers consist of long molecules, each formed as a result of identical groups of atoms called monomers linking together end-to-end to form a long chain. In an unstretched state, the molecules are usually tangled together at random, joined to each other by cross-links which hold the solid together. Where the molecules are folded together in a regular pattern, the polymer in that region is crystalline. When a polymer is stretched, its molecules are straightened out.

see also...

States of Matter; Types of Bonds

Superconductivity

Superconductivity is the complete absence of electrical resistance. A superconductor is a substance with zero electrical resistance. The critical temperature of a superconductor is the temperature at and below which it is super-conducting. The resistivity of a superconducting material falls sharply to zero at its critical temperature as the material is cooled from above its critical temperature to below its critical temperature. Metals and certain alloys and ceramics become superconducting at sufficiently low temperature. A superconducting cable transmits electric current without any heating effect as its resistance is zero. Superconducting magnets are electromagnets consisting of superconducting wires. The very strong magnetic fields created by superconducting magnets are used in magnetic resonance imaging (MRI) systems in hospitals and brain research.

Superconductivity was discovered first in mercury when it was cooled below 4.15 K. Further metals including alloys were found to be superconducting, each with its own critical temperature. Before 1986, the highest critical temperature at 23.3 K was that of an alloy of niobium and germanium. Then superconductivity was discovered at 90 K in a ceramic material. This unexpected discovery was followed by further discoveries using similar materials at higher temperatures. The fact that superconductivity could be achieved by cooling such materials in liquid nitrogen which boils at 77 K led to this class of superconductors being referred to as 'high temperature super-conductors'. At present, the highest critical temperature is about 130 K.

Superconductivity in metals and alloys is known to be due to conduction electrons at large separation, in atomic terms, pairing together. Each pair of electrons, known as a Cooper pair, is a bound state which collides elastically with ions, electrons and other Cooper pairs. Cooper pairs pass through the material without energy loss and therefore with zero resistance.

see also...

Electrical Conduction; Resistance

Temperature

Temperature is the degree of hotness of an object. A temperature scale is defined in terms of fixed points, each fixed point being a well-defined degree of hotness.

★ The Celsius scale of temperature in °C is defined in terms of:
1 ice point, 0 °C, which is the temperature of pure melting ice,
2 steam point, 100 °C, which is the temperature of steam at atmospheric pressure.

★ The absolute scale of temperature, in kelvins (K), is defined in terms of two fixed points which are
1 absolute zero, 0 K, the lowest possible temperature,
2 the triple point of water, 273 K, the temperature at which water, water vapour and steam coexist.

By assigning a value of 273 K to the triple point of water, the interval between ice point and steam point is 100 K, hence absolute temperature in K = temperature in °C + 273.

The First Law of Thermodynamics

states that the heat transfer, ΔQ, to a system is equal to the change of internal energy, ΔU, of the system plus the work done by the system, ΔW.

$$\Delta Q = \Delta U + \Delta W.$$

Internal energy is the energy of the particles of a body due to their random motion or their separation. Work is energy transferred as a result of a force moving its point of application. Heat is energy transferred by any other means.

(i) An adiabatic change is a change which takes place without any heat transfer (i.e. $\Delta Q = 0$). Hence $\Delta U + \Delta W = 0$ for an adiabatic change. Therefore when an object or system of objects does work ΔW in an adiabatic change, its internal energy decreases in accordance with $\Delta U = -\Delta W$.

(ii) An isothermal change is one in which the temperature does not alter. The internal energy of an ideal gas is proportional to its absolute temperature. The work done by an ideal gas when it expands at constant pressure p is equal to $p\,\Delta V$, where ΔV is the change of volume. Therefore for an isothermal change of an ideal gas, $\Delta U = 0$ so $\Delta Q = p\Delta V$.

> ### see also...
> *Entropy; Ideal Gases; States of Matter*

Thermal Properties of Materials

The energy possessed by a substance due to its temperature is referred to as thermal energy. When energy is supplied to a substance to increase its thermal energy, the particles of the substance

★ gain kinetic energy if the temperature of the substance increases, or

★ use the energy supplied to break the bonds between the molecules, if the substance changes state from a solid to a liquid or gas or from a liquid to a gas.

(i) The specific heat capacity of a material is the energy needed to raise the temperature of unit mass of material by one degree.

To raise the temperature of mass m of a substance from T_1 to T_2, the energy needed $\Delta E = mc(T_2 - T_1)$, where c is the specific heat capacity of the material. The unit of c is J kg^{-1} K^{-1} or J mol^{-1}K^{-1}.

(ii) The specific latent heat of a solid or a liquid substance for a given change of state is the energy needed to change the state of unit mass of material, without change of temperature.

To change the state of mass m of a substance at constant temperature, the energy needed $\Delta E = ml$, where l is the specific latent heat of fusion (for melting or solidifying) or vapourisation (for vapourising, boiling or condensing) or sublimation (for a solid which vapourises directly or a vapour which forms a solid without liquefying first) for that substance. The unit of l is J kg^{-1} or J mol^{-1}.

(iii) Thermal expansion is the change of length of a solid when it is heated. When the temperature of a solid or a liquid increases, the particles vibrate with increased amplitude on average which causes the substance to expand. The increase of length is proportional to the initial length and the temperature change. The expansivity α of a substance is defined as the increase of length per unit length per unit temperature rise. The unit of α is K^{-1}. If a length L of a substance is heated from temperature T_1 to temperature T_2, its increase of length $\Delta L = \alpha L (T_2 - T_1)$.

see also...

States of Matter; Temperature

Total Internal Reflection

When a light ray passes from one transparent substance or from air to another transparent substance, its direction changes. This change of direction is known as refraction. Refraction occurs because the speed of light in one substance is different from the speed of light in the other substance.

★ The refractive index, n, of a transparent substance is the ratio of the speed of light in air to the speed of light in the substance. Light travels faster in air than in a transparent solid or liquid.

★ The angle of incidence, i, and the angle of refraction, r, are such that $n_1 \sin i = n_2 \sin r$, where n_1 is the refractive index of the incident substance and n_2 is the refractive index of the other substance.

Critical angle: When a light ray in a transparent substance reaches a boundary with another transparent substance which has a smaller refractive index, total internal reflection of the light ray occurs if the angle of incidence is more than a certain angle, referred to as the critical angle. If the angle of incidence is equal to the critical angle, the angle of refraction is 90°

and the light ray refracts along the boundary. Therefore $n_1 \sin c = n_2 \sin 90$ for a ray at the critical angle; because $\sin 90 = 1$, $\sin c = n_2/n_1$.

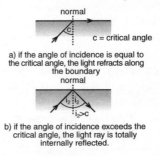

a) if the angle of incidence is equal to the critical angle, the light refracts along the boundary

b) if the angle of incidence exceeds the critical angle, the light ray is totally internally reflected.

Total internal reflection

Application:

1 Telecommunication links use optical fibres to transmit digital signals as light pulses which enter a fibre at one end and undergo total internal reflection each time they reach the fibre surface, emerging at the far end of the fibre.

2 Endoscopes (used to see inside the body) consist of two bundles of optical fibres, one to illuminate the object being observed in the body and the other to guide light scattered by the object to the observer.

see also...

Electromagnetic Waves

Types of Bonds

The electrons in every atom are arranged in shells, each shell capable of holding a certain number of electrons. The innermost shell is capable of holding two electrons, the next shell eight electron and the third shell also eight electrons. The electrons of an atom normally occupy the shells from the innermost outwards. Filled shells represent the lowest possible energy states of an atom. The number of electrons in the outer shell of an atom determines the type of bond it can form with another atom. The atoms of an inert gas do not form bonds because each atom has a full outer shell.

1 Ionic bonds in a crystal hold positive and negative ions together in a regular arrangement referred to as a lattice. Each negative ion is an atom that has gained one or more electrons to form a full outer shell if possible. Each positive ion has lost one or more electrons so as to empty its outer shell.

2 Covalent bonds join atoms together in molecules, in free radicals, and in amorphous solids. Each atom shares one or more of its outer electrons with one or more atoms so each atom has a full outer shell. Each shared pair of electrons is a covalent bond.

3 Metallic bonding occurs in metals where positive metal ions form a regular lattice held together by a 'sea' of free electrons.

4 Van der Waals bonds are weak bonds between neutral atoms or molecules which attract each other as a result of the nucleus of one atom attracting the electrons of another atom.

The Periodic Table of Elements presents the elements in rows, in order of increasing atomic mass from left to right in each row and from top to bottom such that each column consists of a group of elements that share common chemical properties. Each row corresponds to a particular electron shell and each column corresponds to the same number of electrons in the outermost shell. Thus the elements in the same column form the same type of bonds and share common chemical properties.

see also...

States of Matter; Structure of Materials

Uncertainty Principle

The Uncertainty Principle states that the position and momentum of a particle cannot be measured precisely at the same time. The action of measuring one alters the other. For example, the location of an electron may be determined from the direction of scattering of a photon directed at the electron. However, the scattering process would alter the momentum of the electron. In precise terms, the Uncertainty Principle states that the uncertainty in the momentum of a particle $\times$ the uncertainty of its position = $h/2\pi$, where h is the Planck constant. An example of the role of the Uncertainty Principle is provided by β^- decay in which an electron is created in a neutron-rich nucleus and instantly ejected. To confine the electron to the nucleus, the uncertainty in its position would need to be reduced to the diameter of the nucleus which is about 10^{-15} m; therefore, the uncertainty of its momentum, Δp, would be about 10^{-19} kg m s^{-1} (= $h/2\pi \, \Delta x$, where $\Delta x = 10^{-15}$ m and $h = 6.6 \times 10^{-34}$ J s). Its momentum would therefore be at least 10^{-19} kg m s^{-1} which is far too much for it to be held in the nucleus by the electrostatic force of attraction of the protons.

The Uncertainty Principle provides a means of estimating the uncertainty of the energy of a particle or a system of particles over a given time interval. Since no particle can move faster than the speed of light, c, the uncertainty of the position of a particle in time Δt is equal to $c\Delta t$. Hence the uncertainty in its momentum $\Delta p = h/c\Delta t$. For a particle moving at relativistic speeds, it can be shown from $E = mc^2$ that $\Delta E = c\Delta p$. Therefore, the uncertainty of its energy $\Delta E = c\Delta p = h/\Delta t$. This idea is used to explain why an α-particle formed in a nucleus can escape from the nucleus against the strong nuclear force holding it in the nucleus. The particle can borrow energy ΔE to escape provided the time taken to escape, Δt, is less than $h/\Delta E$. The energy needed to escape is an energy barrier which the particle overcomes by borrowing energy from the nucleus for a short period. In effect, the particle 'tunnels' through the barrier. However, if the barrier is too high or too wide, the α particle is unable to escape and the nucleus is therefore stable.

see also...

Quantum Theory; Radioactivity 1

Vectors

★ A vector quantity is any physical quantity that has direction as well as magnitude. Displacement, velocity, acceleration, force, momentum and field strength are examples of vector quantities.

★ A scalar quantity is a physical quantity without directional properties. Examples include distance, speed, mass, energy and power.

A vector quantity may be represented as an arrow of length in proportion to the magnitude of the vector quantity in the appropriate direction. The components of a vector quantity of magnitude F in a direction θ to some specified straight line are $F\cos\theta$ parallel to the line and $F\sin\theta$ perpendicular to the line. If the specified line is the x-axis of a coordinate system, then $F_x = F\cos\theta$ and $F_y = F\sin\theta$. The vector may be written in terms of unit vectors $\mathbf{i}$ and $\mathbf{j}$ along the x- and y-axis respectively, as $\mathbf{F} = (F\cos\theta)\,\mathbf{i} + (F\sin\theta)\,\mathbf{j}$.

The magnitude of a vector F and its direction can be calculated from its perpendicular components F_x and F_y using the rule $F = (F_x^2 + F_y^2)^{1/2}$ and $\tan\theta = F_y / F_x$, where θ is the angle between the vector and the x-axis.

Vector addition:

(i) The parallelogram rule for adding two vectors is an accurate geometrical method of finding the resultant of two vectors. The two vectors are drawn so they form adjacent sides of a parallelogram. The resultant is the diagonal of the parallelogram from the start of one vector to the end of the other vector. The two vectors are joined end-on so the resultant vector is from the start of one vector to the end of the other vector.

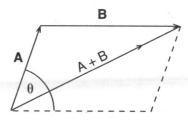

The parallelogram rule

(ii) The cosine rule for adding two vectors **A** and **B** gives the following equation for the magnitude R of the resultant vector **R**: $R^2 = A^2 + B^2 + 2AB\cos\theta$, where θ is the angle between the two vectors.

see also...

Force and Motion; Forces in Equilibrium

Wavemotion 1 – Nature of Waves

Electromagnetic waves, sound waves, seismic waves, water waves and other types of waves have characteristic properties as well as common properties.

(i) Mechanical waves are waves that propagate through a substance due to vibrations of particles of the substance. All types of waves except electromagnetic waves need a substance in which to propagate and are therefore mechanical in nature.

(ii) Electromagnetic waves are electric and magnetic fields that vibrate in phase with each other, propagating through space or a substance. No substance is needed as the fields oscillate regardless of any substance.

A wave propagates due to vibrations in a region creating vibrations in adjacent regions that create vibrations in adjacent regions and so on.

★ Transverse waves are waves in which the vibrations are perpendicular to the direction of propagation of the wave. Examples include all electromagnetic waves, waves on a string and secondary seismic waves.

★ Longitudinal waves are waves in which the vibrations are parallel to the direction of propagation of the wave. Examples include sound waves and primary seismic waves.

Wave measurements: The amplitude of a wave is a measure of how big the wave is. In other words, it is the maximum displacement of a vibrating particle from its equilibrium position. The larger the amplitude of a sound wave, the louder the sound. The larger the amplitude of a water wave, the greater the height of the wave.

★ The wavelength of a wave is the distance along the wave from a wave crest to the next wave crest.

★ The frequency of a wave is the number of wave crests passing a point each second. The unit of frequency is the hertz (Hz), equal to 1 complete cycle per second.

★ The speed of propagation of a wave = its frequency × its wavelength.

see also...

Decibels; Polarisation

Wavemotion 2 – Progressive and Stationary Waves

Progressive or travelling waves are waves that travel through space or through a substance. For mechanical waves, the particles along the direction of propagation of the wave reach maximum displacement at successive instants as each wave crest or wave trough passes. Particles separated by a whole number of wavelengths vibrate in phase with each other.

Stationary or standing waves are produced when two or more travelling waves of the same frequency and amplitude pass through each other. The amplitude of the resultant wave varies with position. Positions of minimum amplitude are referred to as nodes and positions of maximum amplitude are referred to as antinodes. The nodes are formed because the progressive waves are always out of phase by 180° at the positions of the nodes so the progressive waves cancel each other out at these positions. The distance between adjacent nodes is always one half wavelength.

★ Stationary waves can be formed on a vibrating stretched string, with a node at either end. The length of

the string is a whole number of half wavelengths when such a stationary wave is formed. The fundamental mode of vibration is when the string vibrates with only one 'half wavelength' loop along its length.

★ Stationary waves are formed in the air column of an organ pipe when it is made to resonate with sound. A jet of air at one end of the pipe makes a reed vibrate which sends sound waves along the pipe. Some of these sound waves reflect back into the pipe when they reach its open end. Thus sound waves travel through each other inside of the pipe to form a pattern of nodes and antinodes along its length.

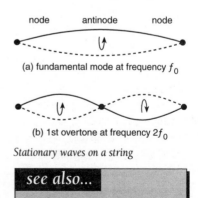

(a) fundamental mode at frequency f_0

(b) 1st overtone at frequency $2f_0$

Stationary waves on a string

see also...

Interference; Wavemotion 1

Wave Particle Duality

Matter particles have a dual nature in that they can behave as waves or as particles according to circumstances. For example, electrons behave as particles when passing through a magnetic field and behave as waves when passing through a thin crystal which acts as a diffraction grating and diffraction is a wave property. The idea that matter particles have a dual nature was first put forward by Louis de Broglie in 1923 as a hypothesis in which he related the momentum of a particle to its so-called de Broglie wavelength, λ, through the equation $\lambda = h/p$, where h is the Planck constant and p is the particle's momentum.

Evidence for wave particle duality was first obtained by George Thomson in 1927 as a result of directing electrons moving at the same speed in a narrow beam at a regular array of atoms in a thin crystal. In this experiment, the beam of electrons is diffracted by the crystal and emerges from the crystal in certain directions only. A photographic film was used to measure the angle of diffraction of each diffracted beam.

The electrons are reflected by each layer of atoms. The electrons from adjacent layers reinforce each other in certain directions only, corresponding to the diffracted orders. For this to be the case, the path difference between the reflected waves from adjacent layers must be a whole number of de Broglie wavelengths. Since the path difference = $2\,d \sin \theta/2$, where d is the layer spacing and θ is the angle of diffraction, then $2\,d \sin \theta/2 = n\lambda$, where n is a whole number. By measuring the angle of diffraction of each diffracted beam, the wavelength can be calculated if d is known. This wavelength value can be checked from the anode potential V of the electron 'gun', using the equation $\lambda = h / (2meV)^{1/2}$, where m is the mass of the electron and e is the charge of the electron. This equation for λ is derived by rearranging the anode gun equation $eV = \frac{1}{2}mv^2$ to give momentum $mv = (2meV)^{1/2}$, then using the de Broglie equation $\lambda = h / mv$ to give $\lambda = h / (2meV)^{1/2}$.

see also...

Quantum Theory

X-rays – Production and Properties

X-rays are electromagnetic waves of wavelengths about 1 nm or less, emitted when fast-moving electrons are decelerated or forced to change direction or when inner-shell vacancies in heavy atoms are refilled. X-rays were discovered in 1895 by Wilhelm Rontgen when he observed that a fluorescent screen glowed when a high voltage current was passed through an evacuated glass bulb. A modern X-ray tube is an evacuated glass tube in which electrons emitted thermionically from a heated filament are attracted onto a metal anode which is at a high positive potential relative to the filament. On impact, the electrons are stopped and lose kinetic energy, releasing X-ray photons in the process.

The X-ray beam contains a continuous spectrum of wavelengths above a minimum value of wavelength. The intensity distribution with wavelength is a smooth curve from the minimum wavelength with intensity spikes superimposed at wavelengths characteristic of the target.

1 The minimum wavelength λ_{min} corresponds to the kinetic energy of a single electron producing a single photon. The gain of kinetic energy is equal to the work done, eV, by the anode potential V, so the photon energy $hc/\lambda_{min} = eV$ (where c is the speed of light), which gives $\lambda_{min} = hc/eV$. The penetrating power of an X-ray beam in a given material depends on the maximum photon energy in the beam which is proportional to the anode potential.

2 The intensity spikes are due to beam electrons striking the target and ejecting inner-shell electrons from the target atoms. When the inner-shell vacancies are refilled by outer electrons of the target atoms, photons are released at wavelengths in the X-ray range. The target metal in an X-ray tube is usually tungsten because it has a high melting point as most of the energy supplied to an X-ray tube is converted to heat. The tungsten target may be set in a copper block as copper has greater thermal conductivity than tungsten.

see also...

Electromagnetic Waves; Photon

X-rays 2 – Use in Medicine

X-rays used for medical imaging need to create sharp images of bones and other high density organs. The patient is placed in the path of the X-ray beam between the X-ray tube and a film cassette in a light-proof container. High density organs or bones absorb X-rays which therefore form an outline of such organs and bones on the film. Low density organs are imaged by filling the organ with a contrast medium which is an X-ray-absorbing substance such as a compound containing barium.

★ The X-rays must therefore originate from as small a spot as possible on the anode otherwise the images are blurred. The anode surface is aligned at an angle of about 70° to the direction of the electrons so that the the effective area from which the X-rays originate is minimised in relation to the electron impact area.

★ In addition, the tube is surrounded by thick lead shielding to prevent X-rays from reaching the operating staff. Lead beam definers are used to limit the X-ray beam to ensure it only passes through the relevant region of the patient.

★ Low energy X-rays are filtered from the beam using a metal plate before the beam passes through the patient. Such low energy photons would be absorbed by low density tissue in the patient, creating unnecessary exposure to the patient.

★ A lead collimator grid is placed between the patient and the film cassette. The grid consisting of a thick lead plate with many narrow parallel holes drilled through it. X-rays scattered by the patient are prevented by this grid from reaching the cassette. Otherwise, these scattered X-rays would expose the shadow regions of the film to X-rays.

see also...

Electromagnetic Waves; X-rays 1

Glossary

Acceleration rate of change of velocity

Activity rate of disintegration of unstable nuclei

Amplitude maximum displacement of an oscillating particle from its equilibrium position

Angular momentum product of the momentum of a particle and the perpendicular distance from a fixed point to its direction of motion

Baryon a composite particle consisting of three quarks

Bond name for any type of force that holds two particles together

Charge there are two types of charge, referred to as positive and negative. Particles that possess the same type of charge repel each other. Particles that possess opposite types of charge attract each other. Charge is quantised in whole number multiples of e, the charge of the electron

Diffusion gradual spread of randomly moving particles in a substance to a uniform distribution

Displacement distance from a fixed point in a given direction

Electron shell the most probable location of the electrons in an atom. The energy of an electron in each shell is constant

Electron volt (eV) 1.6×10^{-19} J, defined as the work done when an electron moves through a p.d. of 1 volt. 1 MeV = 1.6×10^{-13} J

Energy the capacity to change the motion of an object

Force any interaction that can change the motion of an object

Frequency the number of cycles of oscillation of an oscillating object, each cycle being from one extreme to the opposite extreme and back

Hadron any particle that experiences the strong nuclear force

Half-life the time taken for half the number of nuclei of a radioactive isotope to disintegrate

Hole a positive charge carrier in a p-type semiconductor that possesses equal and opposite charge to that of the electron. In effect, a hole is an electron vacancy in an atom in a semiconductor

Intensity energy per second carried by a wave or by radiation incident normally on unit area of a surface

Ion a charged atom. An uncharged atom contains an equal number of electrons and protons. Removal of an electron makes the atom into a positive ion. Addition of an electron makes the atom into a negative ion

Isotope the isotopes of an element are forms of the element which each have the same number of protons but a different number of neutrons in each nucleus

Lepton any particle that experiences the weak nuclear force

Mass a measure of the inertia or resistance to change of motion of an object

Meson a particle consisting of a quark and an antiquark

Momentum the product of the mass and the velocity of an object

Neutron an uncharged particle slightly heavier than the proton. Every atom contains a nucleus which is composed of one or more protons and neutrons

Neutrino an uncharged particle emitted from an unstable nucleus when it decays by emitting a β particle

Phase the fraction of a cycle an oscillating object has passed through in a certain time interval

Proton a positively charged particle which is the nucleus of the lightest atom, the hydrogen atom

Root mean square speed the square root of the mean value of the (speed)^2 of the particles in an ideal gas. Its significance lies in the fact that the mean square speed is proportional to the absolute temperature of the gas

Specific charge the charge/mass value of a charged particle

Speed of light, c the distance travelled by light per unit time. No object can travel faster than the speed of light in a vacuum which is 300 000 kilometres per second

Spin the intrinsic angular momentum of a particle. The spin of a particle is constant and equal to $s(h/2\pi)$, where s is referred to as the spin quantum number which is always a whole number $\times$ $^1/_2$. The spin of an electron is $^1/_2$. Particles with half-integral spin (e.g. $^1/_2$ or $^3/_2$, etc.) are referred to as fermions. Particles with zero or integral spin (e.g. 0 or 1 or 2, etc.) are referred to as bosons

Velocity rate of change of distance is a given direction

Voltage a word used for a potential difference

Wavelength the least distance along a wave between two oscillating particles with the same displacement and velocity at the same instant

Further Reading

Breithaupt, Jim *Key Science Physics for GCSE*, 3rd Edition (Nelson Thornes, 2001)

Breithaupt, Jim *Understanding Physics for Advanced level*, 4th Edition (Nelson Thornes, 2001)

Breithaupt, Jim *Teach Yourself Cosmology* (Hodder and Stoughton, 1999)

Breithaupt, Jim *Teach Yourself 101 Key Ideas in Astronomy* (Hodder and Stoughton, 2000)

Breithaupt, Jim *Einstein: A Beginner's Guide* (Hodder and Stoughton, 2000)

Trigg, George L. *Landmark Experiments in Twentieth Century Physics* (Dover Science Books)